BLUEPRINT READING

NMTBA Shop Practices Series

BASIC SHOP MEASUREMENT

BLUEPRINT READING

SHOP MATH

SHOP THEORY

National Machine Tool
Builders' Association

BLUEPRINT READING

K. O. Hoffman

John Wiley & Sons

New York   Chichester   Brisbane   Toronto   Singapore

Library of Cataloging in Publication Data:

Hoffman, K. E.
  Blueprint reading.
  (NMTBA shop practices series)

  1. Blueprints. I. National Machine Tool
Builders' Association. II. Series.

T379.H572 1983   604.2′5   82-15925
ISBN 0-471-07838-7 (WILEY)
ISBN 0-471-87333-0 (NMTBA)

Printed in the United States of America

10 9 8 7 6 5 4 3 2 1

Foreword

Language students are often surprised when they travel to find that the language they hear around them is not the same as the language they learned in the classroom. Languages such as Chinese have hundreds of dialects or sub-languages. No text on language can represent either the everyday use of the language or be the exclusive source of information on the language.

By comparison, blueprinting is a language that the creator or designer of a part or product uses to describe to the manufacturer or builder of a part how the part is supposed to look on final manufacture.

The NMTBA has more than 400 member companies who use exacting "blueprint language" to communicate within their design and production departments. No matter how precise the blueprinting language is within a company, the same information may have a different meaning in another company.

NMTBA's *Blueprint Reading* introduces the fundamentals of blueprint reading. Certain principles will hold true no matter which blueprint is examined, yet selected elements of a principle or blueprint will vary from company to company and from drafting room to drafting room. Students should learn from this text in the same way that language students learn from their texts. Study and understand the fundamentals of *Blueprint Reading,* keeping in mind that the material may vary slightly from the day-to-day use of blueprints and dimensioning in your shop.

Michael E. Naylon
NMTBA

Contents

BLUEPRINT READING

SECTION I

INTRODUCTION TO BLUEPRINT READING

UNIT 1 # Introduction to Blueprints

Throughout history, the need for accurate communication has caused people to develop many new methods to convey thoughts and ideas. From the first drawings on cave walls to today's written language, communication has played a major role in the development of society.

In industry and manufacturing a principal method of communication is the engineering drawing. Engineering drawings are graphic, picturelike descriptions of objects too detailed to be easily described in words. Drafters, designers, and engineers make these drawings to develop and record their design ideas. These drawings serve as instructions and provide all the information and specifications needed to make an object.

DRAWING REPRODUCTION

Original engineering drawings represent a large investment of both time and money. They also serve as legal documents. Thus, original drawings are normally not used in the shop. Instead, copies are made to prevent damage to the originals.

There are several different ways to copy an original drawing. The oldest form is called the *blueprint.* Blueprints are copies that reproduce an image as white lines on a dark blue background. Today, however, copies are made on either photocopier machines or diazo printing machines. These processes produce copies with either blue, black, or brown lines on a white background.

For industrial purposes, the diazo process is the more common. This process uses a chemically treated, light sensitive paper. The paper is developed with ammonia vapors and a water mixture. Figure 1-1 illustrates the complete diazo process.

After the copies are made, the original is filed for safekeeping. The copies are then sent to the shop. Regardless of the process used or the color of the lines, the terms blueprint, print, and drawing are all shop terms used to describe the reproduced copies of the original engineering drawing.

PARTS OF A BLUEPRINT

There are four basic parts to a blueprint: the body, the title block, the revisions list, and the materials list (Figure 1-2).

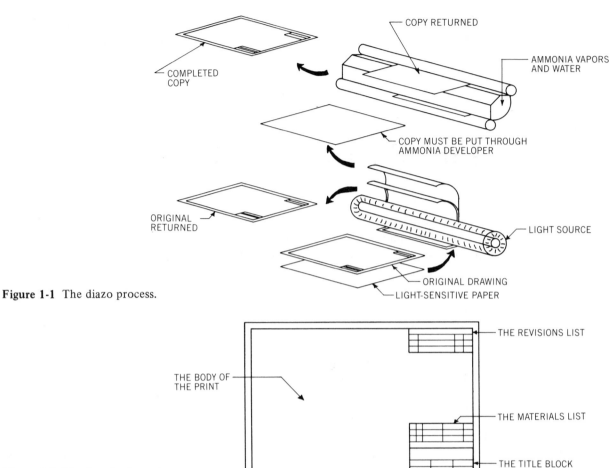

Figure 1-1 The diazo process.

Figure 1-2 The four basic parts of a blueprint.

THE BODY

The body of a blueprint contains the drawn pictures, or *views,* of the object to be made or assembled. These views, along with the size designations, called *dimensions,* and the special instructions, called *notes,* describe the size, shape, and details of the object.

THE TITLE BLOCK

The title block (Figure 1-3) is in the lower right-hand corner of the print. It contains the general information about the part shown in the body. This information includes:

1. *Title.* The title of the print is generally the name of the part or a brief description of the drawn object.

2. *Quantity.* The quantity block indicates the number of the parts to be made. In those cases where the quantity is shown elsewhere on the print, the quantity block will have the word "NOTED."

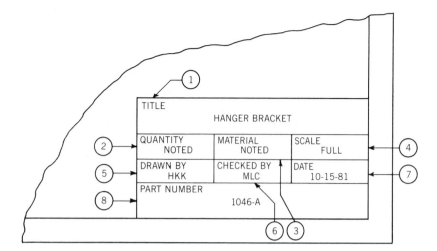

Figure 1-3 The title block.

3. *Material.* The material block lists the specific material used to make the part. Again, if the material specification is shown elsewhere on the print, the word "NOTED" will be used in this block.

4. *Scale.* The scale of the print shows the size relationship between the actual object and the drawn object. A scale of 1:1 means the part in the drawing is the same size as the actual part. Likewise, a scale of 1:2 means the part is drawn half the size of the actual object, and a scale of 2:1 means the drawn object is twice as large as the actual part. When numbers indicate the scale, the first number always refers to the drawing and the second number refers to the actual object. In addition to numbers, the terms FULL, HALF, and QUARTER may also indicate the scale. Table 1-1 shows the standard scales for shop prints. Regardless of the scale, the dimensions shown always refer to the actual size of the object to be made.

5. *Drawn by.* This block shows the initials or name of the drafter who made the original drawing.

6. *Checked by.* This block contains the initials or name of the person who checked the original engineering drawing.

7. *Date.* The date block records the date the original engineering drawing was completed and released for production.

Table 1-1
STANDARD DRAWING SCALES

Eight size = 1:8, 1/8, .125:1.00, or EIGHTH
Quarter size = 1:4, 1/4, .25:1.00, or QUARTER
Half size = 1:2, 1/2, .50:1.00, or HALF

Full size = 1:1, 1 = 1, 1.00:1.00, or FULL

Twice size = 2:1, 2 = 1, or 2.00:1.00
Three times size = 3:1, 3 = 1, or 3.00:1.00
Five times Size = 5:1, 5 = 1, or 5.00:1.00

8. *Part number.* The part number identifies the specific part in the print. This number may occasionally be used as the drawing number.

In addition to the entries already covered, other entries such as sheet number, finish specifications, and heat treatment sometimes appear on specialized prints. The exact form and content of a title block is normally determined by the company making the drawing. For this reason, only the general entries have been discussed here.

THE REVISIONS LIST

The revisions list (Figure 1-4) describes any changes or modifications made to the original drawing. This block is found in the upper right-hand corner of the print. It contains the following information:

1. *Revision number.* The revision number identifies the specific revision. This entry may be either a number or a letter.

2. *Description.* The description entry is a brief note stating the exact change made.

3. *Date.* The date block contains the date the change was made to the original drawing.

4. *Approved.* The approved block records the initials or name of the person who authorized the change.

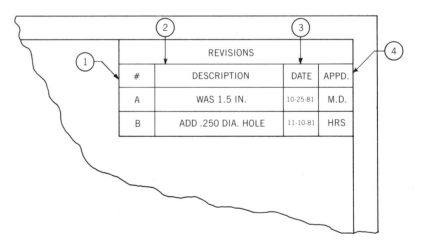

Figure 1-4 The revisions list.

THE MATERIALS LIST

The materials list (Figure 1-5) is generally used on drawings where more than one part is shown in the body. This list contains specific information about the materials or parts used to build the major part or assembly. This list is located on the right side of the print, directly above the title block. It contains the following information:

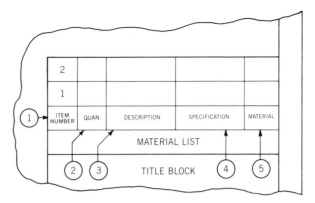

Figure 1-5 The materials list.

1. *Item number.* The item number identifies the specific item in the print. This reference is generally shown as a number and is located next to the part it references.

2. *Quantity.* The quantity block indicates the numbers of each item required for each part or assembly.

3. *Description.* The description is a brief note describing the specific part. In some cases this may be the name of the part or a brief description, such as "ROUND ROD."

4. *Specification.* The specification block records the specific part numbers of commercial parts or the rough stock sizes of parts to be machined.

5. *Material.* The material block lists the specific materials needed to make the part. In cases where a commercial part is used, the note "COMM" is generally used in this block.

Each of the four parts of a blueprint contributes to the overall message of the print. To interpret, or *read,* a blueprint properly, you must study each part to determine the complete meaning.

SELF-TEST

1. What is the name of the oldest kind of drawing reproduction?

2. What is the name of the original drawing?

3. Who makes the original drawings?

4. What processes are most often used to produce blueprints today?

5. Which process is the most commonly used for industrial purposes?

6. What color lines are found on modern blueprints?

7. What three terms are generally used to describe reproduced copies used in the shop?

8. Where are notes and dimensions normally found in a blueprint?

9. If a scale of 2:1 is shown in a blueprint, what size is the drawn object in relation to the actual part?

10. Identify the four parts of a blueprint and show the proper position of each on Figure 1-6, a-d.

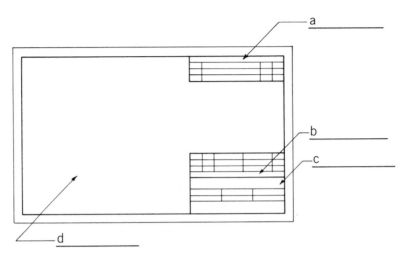

Figure 1-6

Answers to Self Test

1. Blueprint

2. Engineering drawing

3. Drafters, designers, and engineers

4. Photocopy and diazo

5. Diazo

6. Blue, black, and brown

7. Blueprint, print, or drawing

8. In the body

9. The drawn object is twice the size of the actual part.

10. a. Revisions list

    b. Materials list

    c. Title block

    d. Body

UNIT 2   # Interpreting Views

The most common form of drawing for industrial and manufacturing purposes is the *orthographic,* or *multiview.* These drawings present an object in a series of drawings, or views, that represent each side of the object. This allows each side to be shown in its true form and proportion. To understand fully how to read a blueprint, you must first understand how multiview drawings are developed.

PROJECTING VIEWS

The easiest way to visualize the relationship between the drawn views and the actual object is to imagine the object inside a glass box (Figure 2-1). Each view is developed by extending imaginary lines of sight, called *projectors,* from the object to the sides of the box. By extending these projectors, each view is projected to its proper viewing plane on the glass sides of the box.

As shown in Figure 2-2, the projectors are extended from the front of the object to form the front view. Likewise, the projectors extended from the top to form the top view (Figure 2-3). The side view is developed by extending the projectors from the side (Figure 2-4). By extending these projectors from every edge and corner of the object to the viewing plane, the exact boundaries of the views are established. When the box is opened, as shown in Figure 2-5, each of the views revolves into its proper position.

The front, top, and right side views are normally called the three principal views. That is, these three are the views in most multiview drawings. The standard multiview drawing will usually only include the front, top, and right side views. However, to describe some complex objects, there are times when the bottom, rear, or left side views are used in place of, or in addition to, the three principal views. The complete arrangement of all six possible views is shown in Figure 2-6. Regardless of the views selected to describe the part, this is generally the view placement for all multiview drawings.

VIEW SELECTION

Clarity is the first rule a drafter follows when selecting views. The views selected must show the most detail with as few hidden features as possible. For this reason some drawings will have more or less than three views whereas other drawings will use views other than the three principal views.

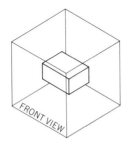

Figure 2-1 Placing the object inside an imaginary glass box.

Figure 2-2 Extending projectors from the front view.

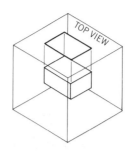

Figure 2-3 Extending projectors from the top view.

Figure 2-4 Extending projectors from the right side view.

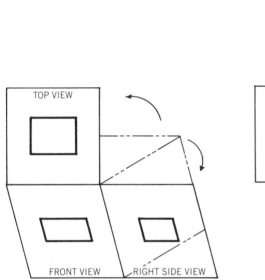

Figure 2-5 Opening the box to form a multiview drawing.

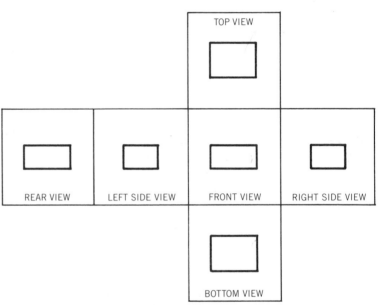

Figure 2-6 Correct arrangement of the views in a multiview drawing.

Figure 2-7, for example, shows six possible views of a typical part. Only two views, the front and right side, are needed to describe this part completely. The other four views are therefore left off the print. Likewise, the part shown in Figure 2-8 also has six possible views. However, this part only requires three views: the front, top, and right side for a complete description.

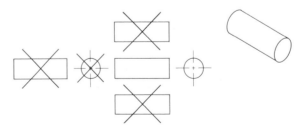

Figure 2-7 Eliminating unnecessary views.

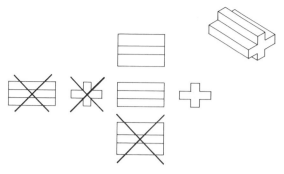

Figure 2-8 Only those views needed for a complete description are shown.

The views used in a multiview drawing must be selected so no detail is left out and repeated details are kept to a minimum. Only the part itself can finally determine the number and position of the views that should be used. The important point to remember here is the position of each view and its relationship to the other views.

VISUALIZATION   The act of visualizing is a key to reading a blueprint. Visualizing is the process of developing a three-dimensional mental picture of the part from the two-dimensional views you see in the blueprint.

Each view in a multiview drawing shows each side of an object in a two-dimensional form. That is, the top and bottom views show the width and depth; the front and rear views show width and height; and the left and right side views show the depth and height of the object (Figure 2-9). To interpret the meaning of a blueprint properly, study each view. Relate it to the other views. Try to develop a mental picture of how the finished part should appear.

Try the steps one by one. First, study the front view of the part in Figure 2-10. This will give you a general idea of the overall form of the part. The front views shows the part to be a basic "L" shape with a long horizontal member

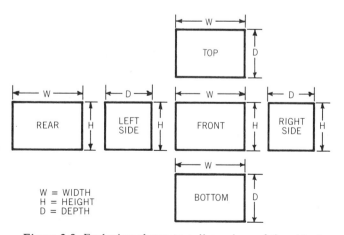

Figure 2-9 Each view shows two dimensions of the object.

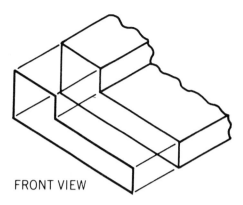

FRONT VIEW

Figure 2-10 Visualizing the front view.

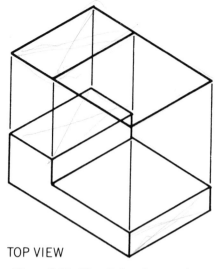

TOP VIEW

Figure 2-11 Visualizing the top view.

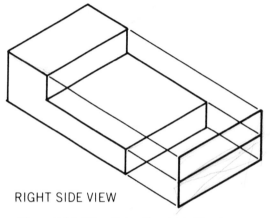

RIGHT SIDE VIEW

Figure 2-12 Visualizing the right side view to determine the complete shape of the actual object.

and a short vertical member. Next, add the top view to the mental picture you formed from the front view; note the square end of the horizontal member and the depth of the part (Figure 2-11). Finally, add the right side view to your mental picture (Figure 2-12). A picture of the actual object should begin to form in your mind. By following this simple process, you can quickly and accurately visualize almost any part.

SKETCHING

Another method to help you visualize a part is sketching. To visualize how the three views should appear, simply use graph paper to sketch each side of the object as shown in Figure 2-13. The part is first measured, and the front view is sketched on the graph paper (Figure 2-14). The top and right side views are then added to the sketch by extending the boundary lines of the front view (Figure 2-15). The final step is to erase your construction lines and darken the lines that show the three views of the part (Figure 2-16).

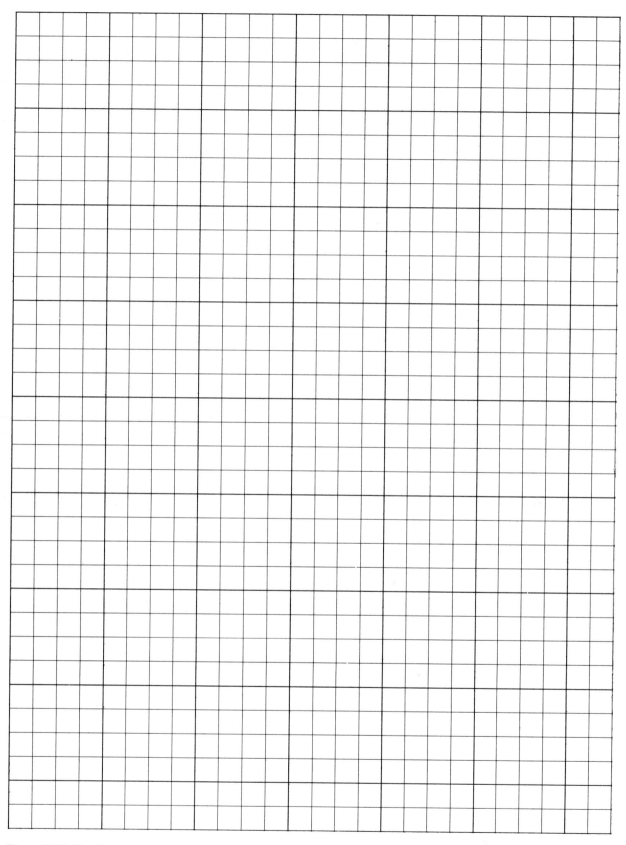

Figure 2-13 Graph paper.

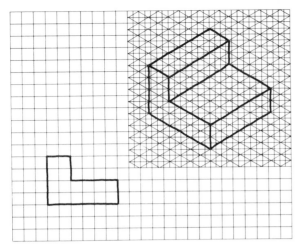
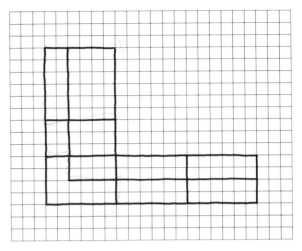

Figure 2-14 Sketching the front view.

Figure 2-15 Constructing the top and right side views.

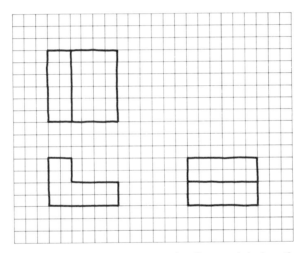

Figure 2-16 Erase the construction lines and darken the
lines that show the three views to finish the sketch.

To construct a pictorial view of an object from a three-view print, you'll need to use another type of graph paper. This graph paper (Figure 2-17) is called *isometric graph paper*. It can be purchased in most office supply stores. When using this paper, simply note the overall sizes of the part and sketch in the rectangular form of the object (Figure 2-18). Once this is done, find the sizes of the details, such as holes, slots, or ledges. Add these to the rectangular outline (Figure 2-19). The final step is to erase all your construction lines and darken the lines that show the pictorial view of the part (Figure 2-20).

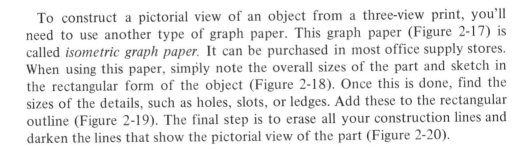

SELF TEST

1. What type of drawing is the most common in manufacturing?

2. What are the imaginary lines of sight called?

3. Which three views are called the principal views?

Figure 2-17 Isometric graph paper.

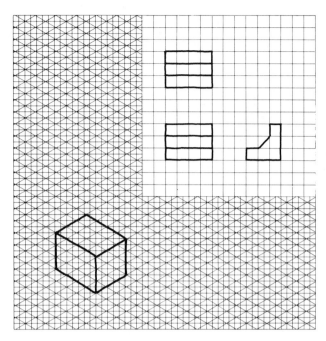

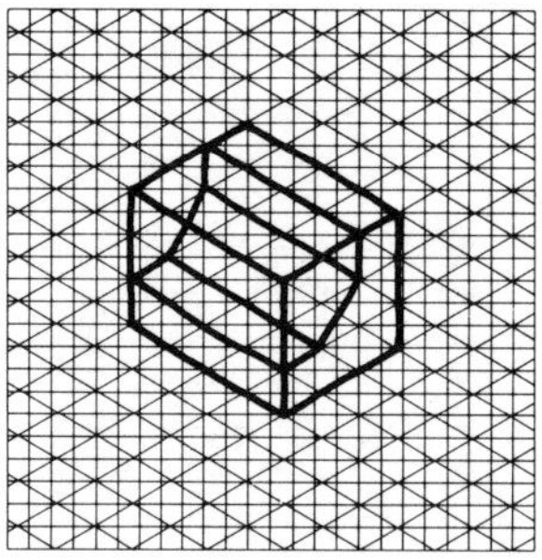

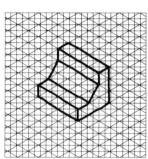

Figure 2-19
Completing the details.

Figure 2-20 Darken the
lines to finish the sketch.

Figure 2-18 Constructing the rough outline of the object.

4. Identify the position of the views shown in Figure 2-21.

5. What determines the number of views used to describe a part?

6. Which views show the following?

Height:	a. Top
Depth:	b. Bottom
Width:	c. Left side
	d. Right side
	e. Front
	f. Rear

7. When visualizing a part, which view should you study first?

8. Why are engineering drawings drawn as a series of views?

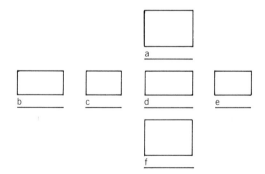

Figure 2-21

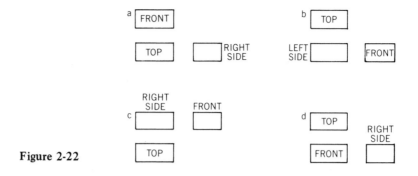

Figure 2-22

9. Identify the correct view placement for a standard three-view drawing in Figure 2-22.

10. Match the multiview drawings in Figure 2-23a to the objects they represent in Figure 2-23b.

11. Make a three-view sketch of the part shown in Figure 2-24.

12. Make a pictorial sketch of the part shown in the three-view drawing in Figure 2-25.

Answers to Self Test

1. Orthographic or multiview

2. Projectors

3. Top, front, and right side

4. a. Top   d. Front
   b. Rear   e. Right side
   c. Left side   f. Bottom

5. The complexity of the part

6. Height: c, d, e, f
   Depth: a, b, c, d
   Width: a, b, e, f

7. Front view

8. To show each side in its true form and proportion

9. d

10. a. 5   f. 4   11. See Figure 2-26.
    b. 9   g. 6   12. See Figure 2-27.
    c. 2   h. 1
    d. 10   i. 8
    e. 3   j. 7

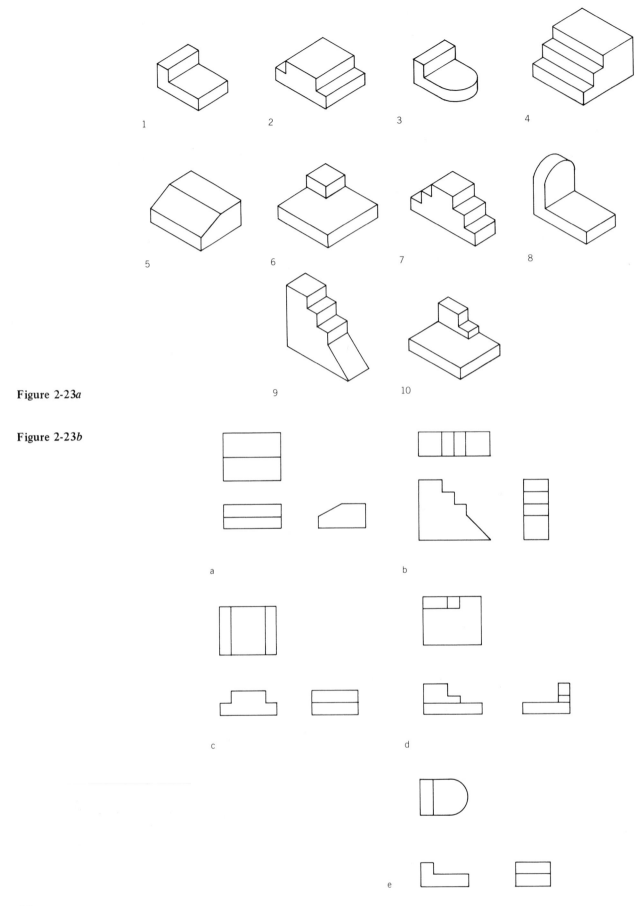

Figure 2-23a

Figure 2-23b

18

Figure 2-23b
(continued)

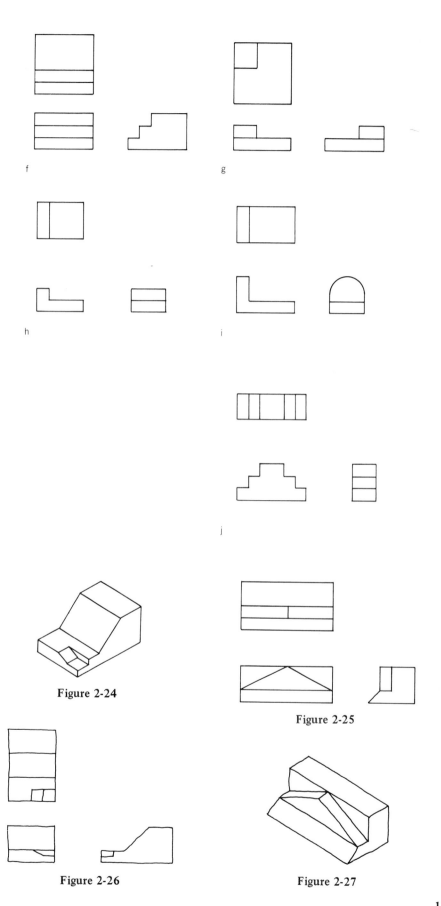

f

g

h

i

j

Figure 2-24

Figure 2-25

Figure 2-26

Figure 2-27

UNIT 3   # Alphabet of Lines

The *alphabet of lines* is a system that gives specific meaning to the individual lines in an engineering drawing. To interpret the meaning of any print, you must be able to identify each line and know exactly what each line means.

TYPES OF LINES   There are ten different types of lines in this basic alphabet. These lines are shown in Figure 3-1 and listed here.

1. Object lines.

2. Hidden lines.

3. Center lines.

4. Phantom lines.

5. Dimension lines.

6. Extension lines.

7. Leader lines.

8. Cutting plane lines.

9. Section lines.

10. Break lines.

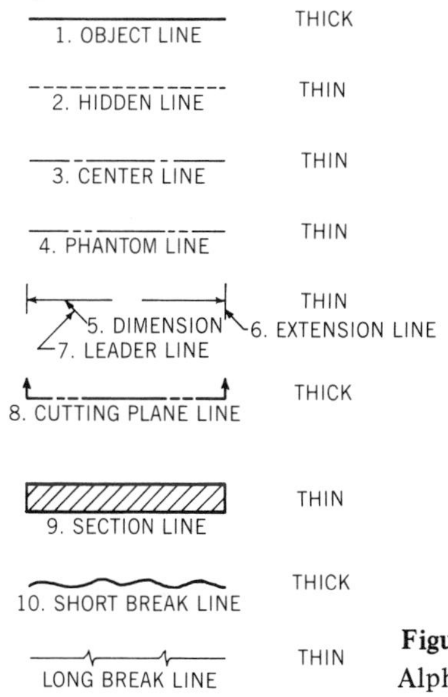

1. OBJECT LINE   THICK

2. HIDDEN LINE   THIN

3. CENTER LINE   THIN

4. PHANTOM LINE   THIN

5. DIMENSION   THIN
7. LEADER LINE   6. EXTENSION LINE

8. CUTTING PLANE LINE   THICK

9. SECTION LINE   THIN

10. SHORT BREAK LINE   THICK

LONG BREAK LINE   THIN

Figure 3-1
Alphabet of lines.

OBJECT LINES
Object lines (Figure 3-2) are the most commonly used lines on engineering drawings. These lines show the visible positions of a part, such as the edges, surfaces, and corners. Since object lines are solid, thick lines, they allow the part to be clearly visible on the drawing.

HIDDEN LINES
Hidden lines (Figure 3-3) are constructed as thin, evenly spaced dashes. These lines are used to show edges, surfaces, and other details that are not directly visible in a particular view.

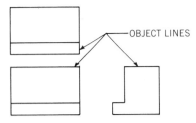

Figure 3-2 Object lines.

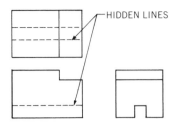

Figure 3-3 Hidden lines.

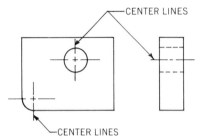

Figure 3-4 Center lines.

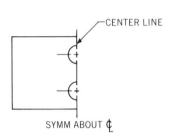

Figure 3-5 Showing symmetry with center lines.

CENTER LINES

Center lines (Figure 3-4) are thin lines with alternating long and short dashes. These lines may appear in many places and serve several functions on a print. Center lines show the center positions of holes, arcs, and radii; they indicate the center axes of parts; and they show symmetry.

When center lines are used to show symmetry, or equality on both sides of the center line, the note SYMM ABOUT ₵ is normally used (Figure 3-5). The symbol ₵ means center line.

PHANTOM LINES

Phantom lines are thin lines with alternating long dashes separated with two short dashes. These lines are used in many ways. As shown in Figure 3-6, phantom lines may show alternate positions of moving parts. Here the lever is shown in one position with object lines. Its other position is shown with phantom lines. The path of motion is normally shown with a center line. Phantom lines also show the position and relationship of adjacent parts, and they eliminate unnecessary repeated details (Figure 3-7).

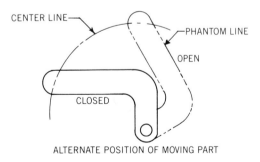

Figure 3-6 Phantom lines.

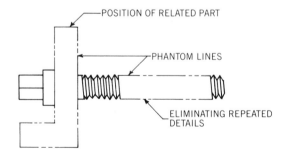

Figure 3-7 Various uses of phantom lines.

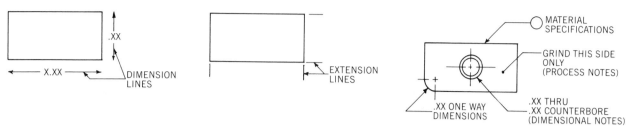

Figure 3-8 Dimension lines.   **Figure 3-9** Extension lines.   **Figure 3-10** Leader lines.

DIMENSION LINES

Dimension lines (Figure 3-8) are thin, solid lines showing the direction and extent of dimensions. These lines are drawn with an arrowhead at each end and an open space near the middle where the dimensional value is inserted.

EXTENSION LINES

Extension lines (Figure 3-9) are thin lines that show the limits of dimension lines. These lines are drawn perpendicular to the dimension lines and close to the edges they limit. The length of the extension line is determined by the number of dimensions they limit. In most cases the extension line will extend slightly past the last dimension line it limits.

LEADER LINES

Leader lines are thin, solid lines with an arrowhead at one end and a bend at the other end (Figure 3-10). Leader lines, or leaders, are used to show notes such as material specifications and processes, one-way dimensions, or dimensional notes. When a note refers to a surface rather than to an edge, a dot is used on the end of the leader line instead of an arrowhead.

CUTTING PLANE LINES

Cutting plane lines (Figure 3-11) are thick lines with long dashes alternating with two short dashes. These lines show the path of an imaginary cut made through the part to form a sectional view. The arrowheads on each end of the cutting plane line show the viewing direction of the sectional view. Figure 3-12 shows an offset cutting plane line. This line shows a sectional view that contains the most detail and explains the internal details of the holes.

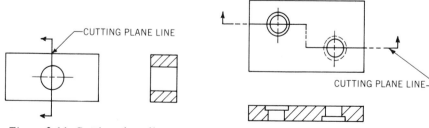

Figure 3-11 Cutting plane lines.   **Figure 3-12** Offset cutting plane.

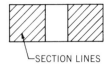

└─SECTION LINES

Figure 3-13 Section lines.

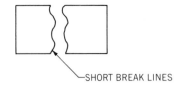
└─SHORT BREAK LINES

Figure 3-14 Short break lines.

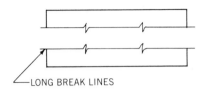
└─LONG BREAK LINES

Figure 3-15 Long break lines.

SECTION LINES

Section lines (Figure 3-13) are thin, solid lines used with cutting plane lines. They show the exposed edges as they would appear if the part were actually cut. These lines are generally drawn at a 45° angle and will sometimes be used to identify the material used to make the part.

BREAK LINES

Break lines are used to shorten objects that are too long to fit the drawing sheet. When break lines are used, you can assume that the portion removed is identical to that shown on either side of the break. The two primary types of break lines are *short break lines* and *long break lines.* Short break lines (Figure 3-14) are thick, solid, wavy lines that are drawn freehand. Long break lines (Figure 3-15) are thin, solid lines with zig zags to indicate the break.

SELF TEST

1. Sketch the following lines:

   a. Hidden line   g. Long break line

   b. Phantom line   h. Center line

   c. Short break line   i. Cutting plane line

   d. Extension line   j. Leader line

   e. Object line   k. Section line

   f. Dimension line

2. Name four places a center line could be used.

3. List three possible applications of phantom lines.

4. What line shows the invisible edges of a part?

5. What line shows the visible edges of a part?

6. Which line is used to show one-way dimensions?

7. What line is always used to show the exposed edges of a sectional view?

8. Which line is normally used with a dimension line?

9. Which break line uses a thin line? Which uses a thick line?

10. What is shown by the arrowheads on a cutting plane line?

**Answers to
Self Test**

1.

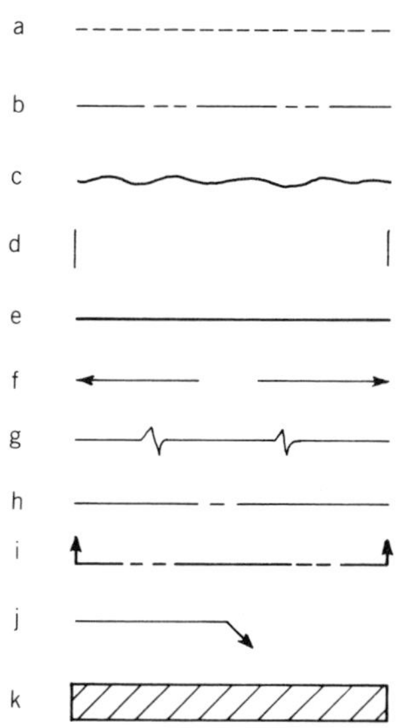

a ---------------

b ——— — — ——— — — ———

c 〜〜〜〜〜〜

d | |

e ————————

f ⟵——————   ————————⟶

g ——⋀⋀——⋀⋀————

h ——— — — ———

i ↥———— — — ————↥

j ——————⟍

k ▱▱▱▱▱▱   **Figure 3-16**

2. Any four are correct:

a. Center positions of holes

b. Center positions of radii

c. Part axes

d. Part symmetry

e. Paths of motion

3. a. Positions of adjacent parts

b. Alternate positions of moving parts

c. Elimination of repeated details

4. Hidden line

5. Object line

6. Leader line

7. Section line

8. Extension line

9. Long break line, short break line

10. Viewing direction

WORKING WITH BLUEPRINTS

UNIT 4   # Detail and Assembly Prints

Multiview drawings used in the shop are normally referred to as *working prints*. These prints can be divided into two general catagories: *detail prints* and *assembly prints*. To make or assamble any parts properly, you must first understand these prints and how they are used.

DETAIL PRINTS   Detail prints completely describe a single part. These prints include all the information and specifications needed to make the part in the drawing. In most cases, detail prints show only a single part on each print. However, there are times when several detail drawings will appear on one large sheet. In either case, the intent of a detail print is the same: to provide enough information and instructions to complete the part shown.

The basic detail print is simply a multiview drawing with enough views to describe the part completely. Most detail prints contain three views; but there are times when two view or one view prints may be used. Here again, the part itself determines the number of views. As a general rule, drafters will draw the minimum number of views necessary to describe the object completely.

THREE-VIEW DETAIL PRINTS
Three-view detail prints (Figure 4-1) are the most common form of print used in the shop. These prints, like other multiview drawings, use different views to describe the part. The views generally used are the front, top, and right side. Only occasionally, with very detailed parts, will more than three views be used.

TWO-VIEW DETAIL PRINTS
Two-view detail prints (Figure 4-2) are used for parts that have symmetrical features or when a third view would repeat one of the other views (Figure 4-3). Although any two views may be used, the front and right side views are the most common.

ONE-VIEW DETAIL PRINTS
The one-view detail print is actually a modified form of the two-view print. One-view prints use a note to replace a drawn view when a separate view would add little to the drawing. As shown in Figure 4-4, the simple note, .XX THICK,

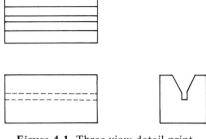

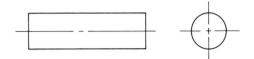

Figure 4-2 Two-view detail print.

Figure 4-1 Three-view detail print.

Figure 4-4 Eliminating a view with a note.

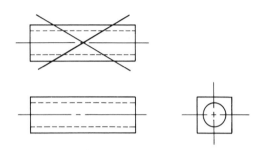

Figure 4-3 Eliminating a repeated view.

Figure 4-5 Using diameter symbols or abbreviations to replace a view.

eliminates the need for another view and simplifies the drawing. When the part is cylindrical, or round, the abbreviation "DIA," or the diameter symbol "ϕ" is enough to replace a view (Figure 4-5).

ASSEMBLY PRINTS

Assembly prints are used to show the position and functional relationships of parts in an assembly. Like detail prints, assembly prints are also multiview prints. They show the assembled unit in a series of separate views. In simple assemblies there may only be one view; more complex units may require two or three views for a complete description. Assembly prints are made in several variations, such as complete assembly, subassembly, and pictorial assembly. Although there are several variations, all assembly drawings can be divided into two general catagories: *unit assembly prints* and *detail assembly prints.*

UNIT ASSEMBLY PRINTS

Unit assembly prints show complex units in their final assembled form. The main purpose of a unit assembly print is to show the position and functional arrangement of each part in the assembled unit. As a general rule, these prints do not show the individual sizes of each part in the assembly. Rather, a unit assembly print, when dimensioned, will normally show the overall size of the completed unit.

Each part in a unit assembly print is identified by a reference callout. The callout is referenced to an entry in the materials list (Figure 4-6). This reference

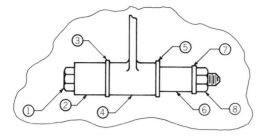

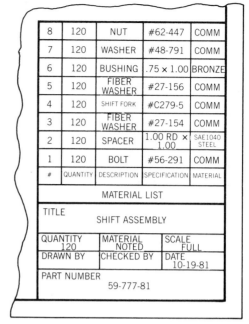

8	120	NUT	#62-447	COMM
7	120	WASHER	#48-791	COMM
6	120	BUSHING	.75 × 1.00	BRONZE
5	120	FIBER WASHER	#27-156	COMM
4	120	SHIFT FORK	#C279-5	COMM
3	120	FIBER WASHER	#27-154	COMM
2	120	SPACER	1.00 RD × 1.00	SAE1040 STEEL
1	120	BOLT	#56-291	COMM
#	QUANTITY	DESCRIPTION	SPECIFICATION	MATERIAL

MATERIAL LIST

TITLE	SHIFT ASSEMBLY	
QUANTITY 120	MATERIAL NOTED	SCALE FULL
DRAWN BY	CHECKED BY	DATE 10-19-81
PART NUMBER 59-777-81		

Figure 4-6 Unit assembly prints use callouts to reference parts in the materials list.

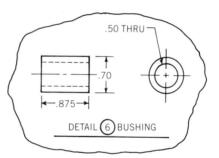

Figure 4-7 Using a callout to identify the detail print.

callout is generally a number, but it may also be a letter. In cases where the unit assembly print is part of a package, or set, of detail prints, the reference callout can also identify the specific detail print (Figure 4-7). Regardless of how it is used, the overall purpose of a unit assembly print is the same: to identify the parts of an assembly and to show how they go together.

DETAIL ASSEMBLY PRINTS

Detail assembly prints, or working assembly prints, show both the position and the size of each part in an assembled unit. Since all the required information can be placed on a single print, this type of print is sometimes preferred for very simple assemblies.

The two main forms of detail assembly prints show the size of each part either by dimensioning the assembly print (Figure 4-8) or by using detail drawings along with the assembly drawing (Figure 4-9). In either case, the print will include all the information required to make each part and to assemble the complete unit.

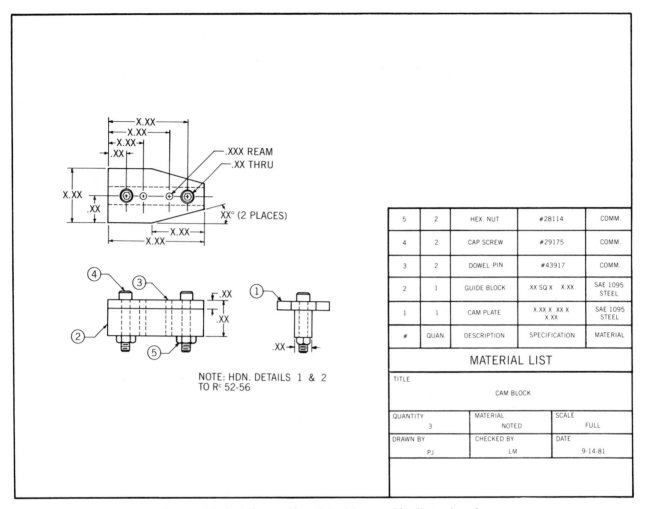

Figure 4-8 Detail assembly print with assembly dimensioned.

5	2	HEX. NUT	#28114	COMM.
4	2	CAP SCREW	#29175	COMM.
3	2	DOWEL PIN	#43917	COMM.
2	1	GUIDE BLOCK	.XX SQ X X.XX	SAE 1095 STEEL
1	1	CAM PLATE	X.XX X .XX X X.XX	SAE 1095 STEEL
#	QUAN.	DESCRIPTION	SPECIFICATION	MATERIAL

MATERIAL LIST

TITLE			
	CAM BLOCK		
QUANTITY 3	MATERIAL NOTED	SCALE FULL	
DRAWN BY PJ	CHECKED BY LM	DATE 9-14-81	

NOTE: HDN. DETAILS 1 & 2
TO Rᶜ 52-56

SELF TEST

1. What is the name of multiview drawings that are used in the shop?

2. What two general categories are used to classify multiview drawings that are used in the shop?

3. What kind of print is normally used for a single part?

4. How many views are normally used with this kind of print?

5. What replaces a view in a one-view multiview print?

6. What do the abbreviations "DIA" and "ϕ" mean?

7. How many general catagories are used to classify prints that describe assembled units? What are they called?

8. Which prints show only overall dimensions?

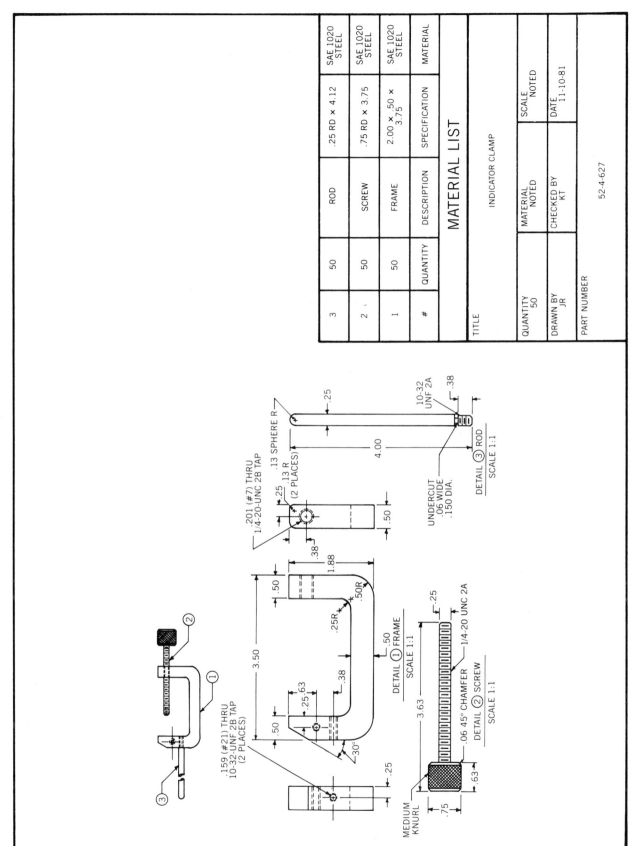

#	QUANTITY	DESCRIPTION	SPECIFICATION	MATERIAL
3	50	ROD	.25 RD × 4.12	SAE 1020 STEEL
2	50	SCREW	.75 RD × 3.75	SAE 1020 STEEL
1	50	FRAME	2.00 × .50 × 3.75	SAE 1020 STEEL

MATERIAL LIST

TITLE	INDICATOR CLAMP	
QUANTITY 50	MATERIAL NOTED	SCALE NOTED
DRAWN BY JR	CHECKED BY KT	DATE 11-10-81
PART NUMBER	52-4-627	

.25

10-32 UNF 2A

.38

.13 SPHERE R

4.00

UNDERCUT .06 WIDE .150 DIA.

DETAIL ③ ROD
SCALE 1:1

.201 (#7) THRU 1/4-20-UNC 2B TAP

.25 .13 R (2 PLACES)

.38

.50

1.88

.50R

.25R

.50

.50

DETAIL ① FRAME
SCALE 1:1

.25

1/4-20 UNC 2A

3.63

.06 45° CHAMFER

DETAIL ② SCREW
SCALE 1:1

MEDIUM KNURL

.63

.75

30°

.50

.25 .63

.38

3.50

.159 (#21) THRU 10-32-UNF 2B TAP (2 PLACES)

.25

②

①

③

Figure 4-9 Detail assembly print with separate details dimensioned.

33

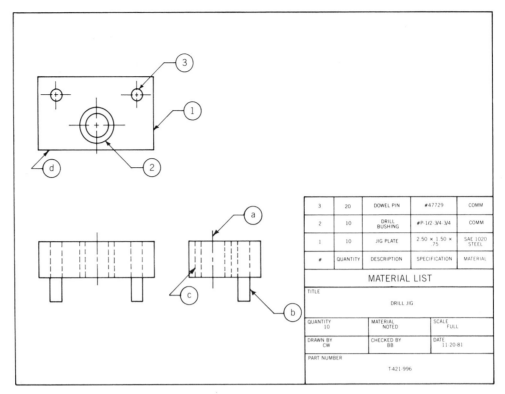

Figure 4-10

9. Which prints show all the information needed to make and assemble a complete unit?

10. What type of print is shown in Figure 4-10?

11. What is the part name?

12. What is the part number?

13. Identify the views shown.

14. Name parts 1-3.

15. How many parts are required?

16. What material is specified for part 1?

17. What is the scale of the print?

18. If this print were completely dimensioned, what type of print would it be?

19. If only part 2 were drawn, what type of print would this be?

20. Identify lines a-d.

**Answers to
Self Test**

1. Working prints

2. Detail and assembly prints

3. Detail print

4. Three views

5. A note

6. Diameter

7. Two. Unit assembly prints and detail (working) assembly prints

8. Unit assembly prints

9. Detail assembly prints

10. Unit assembly

11. Drill jig

12. T-421-996

13. Top, front, and right side.

14. 1. Jig plate

    2. Drill bushing

    3. Dowel pin

15. 10

16. SAE 1020 steel

17. Full

18. Detail assembly print

19. Detail print

20. a. Center line

    b. Object line

    c. Hidden line

    d. Object line

Special Views

In addition to the standard views found in multiview prints, there are several special views for special situations. These views include sectional views, auxiliary views, and detail views. To interpret a complex print properly, you must understand these views and the message they convey.

SECTIONAL VIEWS

The external shape of an object is shown in a multiview print with object lines. Internal features, such as holes and slots, are shown with hidden lines (Figure 5-1). But when the shape of the internal features are very detailed or complex, hidden lines will tend to confuse the meaning of the internal feature. The internal features of the part shown in Figure 5-2 are shown using hidden lines. These hidden lines are very difficult to understand. The true shape of the internal feature is not easily visualized. But if a sectional view is used (Figure 5-3), the meaning of the hidden lines becomes clear and the part is easier to visualize. Drafters use sectional views to show the true shape of internal features and to reduce the chance of misinterpretation.

Sectional views are indicated on a print with cutting plane lines. Cutting plane lines show the location and path of an imaginary cut through the part. This imaginary cut indicates the position of the sectional view (Figure 5-4). To understand how a cutting plane line is related to the sectional view, simply imagine the part actually being cut along the cutting plane line. The sectional view that is formed represents the way the part would appear if it were actually cut.

The placement of the sectional view in the print is directly related to the viewing direction. This viewing direction is determined by the direction of the arrowheads on the ends of the cutting plane line. For example, if the cutting plane line were positioned as shown in Figure 5-5, the sectional view would be drawn as a front view. Likewise, if the cutting plane line were positioned as shown in Figure 5-6, the sectional view would be drawn as a right side view.

Sectional views are identified by letters that refer to a specific cutting plane line (Figure 5-7). The cutting plane line is identified by a letter at each end, and the sectional view is labeled "SECTION AA". When several sectional views appear on a single print, the letters continue from AA, BB, CC, DD, and so on through the alphabet.

The exposed edges of a sectional view are indicated on a print by section lines. These lines show the surfaces that would be exposed if the part were actually cut. Section lines are generally shown as thin lines drawn at 45° angles (Figure 5-8). In some cases, section lines identify the specific materials as well as showing the edges of a sectional view. Figure 5-9 shows the standard forms of section lines and the materials they represent. Unless otherwise specified, the plain 45° section lines are used for general purpose sectioning and do not always mean the material is cast iron.

 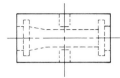

Figure 5-1 Hidden lines used to describe simple internal details.

Figure 5-2 Hidden lines used to describe complex internal details.

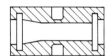 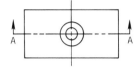 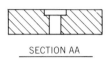

Figure 5-3 A sectional view clarifies the meaning of hidden lines.

Figure 5-4 Using cutting plane lines to indicate a sectional view.

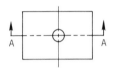

Figure 5-5 Arrowheads determine the viewing direction of this part to be a front view.

Figure 5-6 The viewing direction of this part is from the right side.

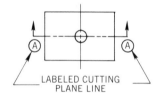 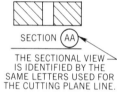 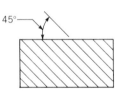

Figure 5-7 Labeling cutting plane lines and sectional views.

Figure 5-8 Section lines.

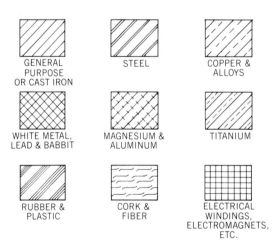

Figure 5-9 Section lines used to identify materials.

TYPES OF SECTIONAL VIEWS

Sectional views are drawn in several different ways. Depending on what internal details must be shown, sectional views can be drawn as full sections, half sections, offset sections, removed sections, broken sections, or revolved sections.

FULL SECTIONS

Full sectional views are used to show the complete internal detail of a part. With these sectional views, the cutting plane line is drawn completely through the part. In cases where the part is round, or where a hole is sectioned, the cutting plane line is drawn in place of the center line, along the center axis of the part (Figure 5-10).

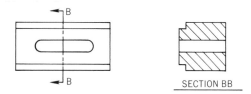

Figure 5-10 Full sectional view.

HALF SECTIONS

Half sections show the internal details of symmetrical parts. Since both sides of a symmetrical feature, such as holes, are the same, only one side needs to be sectioned for a complete description (Figure 5-11). Half sectional views also have the advantage of showing both internal and external details in the same view. When half section views are used, it is automatically assumed that the area not shown, both internal and external, is identical to the area shown.

One other point to look for in some half section views is the cutting plane line. In some cases, the cutting plane line will have only one arrowhead. This means the viewing direction is parallel to the cutting plane line (Figure 5-12).

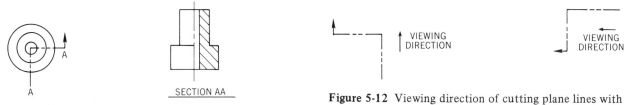

Figure 5-11 Half sectional view.

Figure 5-12 Viewing direction of cutting plane lines with only one arrowhead.

OFFSET SECTIONS

Offset sections are drawn to show the most detail with a single section. Rather than using several short cutting plane lines, an offset section view uses a single cutting plane line positioned to show all the required internal details in a single offset section (Figure 5-13).

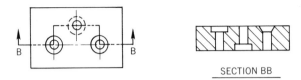

Figure 5-13 Offset sectional view.

REMOVED SECTIONS

Removed sections are sectional views that are projected directly from the area of the part they represent (Figure 5-14). These sectional views frequently show details such as cross sectional views of spokes, levers, or similar parts.

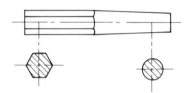

Figure 5-14 Removed sectional view.

Figure 5-15 Broken sectional view.

BROKEN SECTIONS

Broken sections show specific, limited areas of a part as sectional views. These sections are mainly used where other forms of sectional views would show more internal detail than is required, (Figure 5-15). When broken sections are shown in a print, a cutting plane line will not indicate the section. Instead, a simple short break line is used to show the edges of the section.

Figure 5-16 Revolved sectional view.

REVOLVED SECTIONS

Revolved sections show sectional views of simple parts. These sections generally indicate the basic external shape of the sectioned area rather than any internal detail. This permits details such as ribs, spokes, and similar details to be described completely. Revolved sections may appear on a blueprint as either superimposed onto the area sectioned or in a broken-out area, as shown in Figure 5-16.

AUXILIARY VIEWS

Auxiliary views indicate the size and shape of surfaces that cannot be accurately shown in any of the three principal views. In Figure 5-17, the inclined, or angled, surface of the part appears distorted and is not drawn in its true form or proportion. To show this surface properly, an auxiliary view, as in Figure 5-18, may be used.

Auxiliary views are constructed at right angles to the surfaces they represent. That is, an auxiliary view is drawn perpendicular to the principal view it describes. These views are identified by the principal views with which they are aligned. In Figure 5-19, a front auxiliary view is aligned with the front view, a

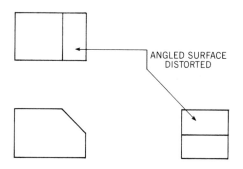

Figure 5-17 An inclined surface shown in the three principal views makes the surface appear distorted.

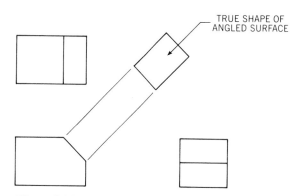

Figure 5-18 An auxiliary view clarifies the shape and form of the inclined surface.

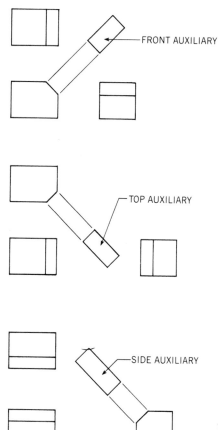

Figure 5-19 Auxiliary views are identified by the principal view with which they are aligned.

top auxiliary view is aligned with the top view, and the side auxiliary is drawn perpendicular to the side view.

The two main types of auxiliary views are the *primary auxiliary view* and the *secondary auxiliary view*. Primary auxiliary views are those auxiliary views that are developed from one of the principal views (Figure 5-20). Secondary auxiliary views are those developed from a primary auxiliary view (Figure 5-21).

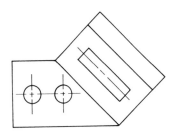

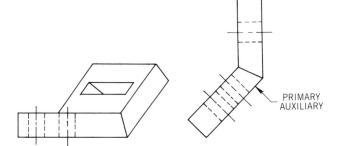

Figure 5-20 Primary auxiliary view.

Figure 5-21 Secondary auxiliary view.

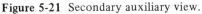

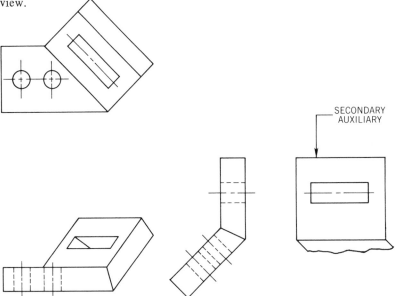

For the most part, primary auxiliary views are the most common. Only rarely will a part require a secondary auxiliary view.

In most cases, when auxiliary views are used, only the inclined surface is shown in the auxiliary view. To reduce the chance of confusion, the other areas of the part are not drawn. In special situations auxiliary views may also be shown as sectional views. The view is then refered to as an *auxiliary sectional view* (Figure 5-22).

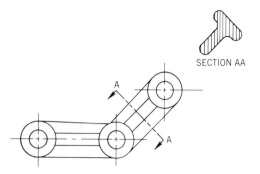

Figure 5-22 Auxiliary sectional view.

DETAIL VIEWS

Detail views show areas of a part that require special attention. These views allow a protion of the part to be drawn in greater detail for clarity or to show special instructions. The two primary types of detailed views are *enlarged detail views* and *sectional detail views.*

ENLARGED DETAIL VIEWS

Enlarged detail views call attention to areas of a part that cannot be accurately shown in any of the principle views. For example, the keyway in the part shown in Figure 5-23 is drawn too small to be accurately interpreted. When this area of the part is shown as an enlarged detail view, the keyway is easier to see and can be accurately dimensioned.

When detail views are shown on a print, the area to be shown in the enlarged view will have the note "DETAIL A" and the scale of the enlargement line (Figure 5-24). A letter is used to identify the exact detail view. The enlarged view will have the not "DETAIL A" and the scale of the enlargement drawn under the view.

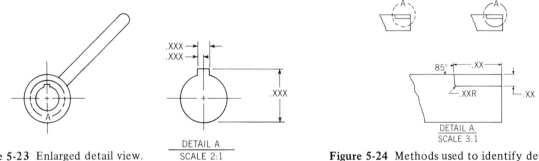

Figure 5-23 Enlarged detail view.

Figure 5-24 Methods used to identify detail views.

SECTIONAL DETAIL VIEWS

Sectional detail views are used to show enlarged internal details of a part that cannot be clearly shown in any of the principal views. As shown in Figure 5-25, sectional detail views are shown on a print in the same way as enlarged detail views. In most cases these sectional views appear as broken sections. In other words, only the required area is drawn as a sectional view.

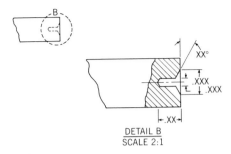

Figure 5-25 Sectional detail views.

SELF TEST

1. What is the main reason for using sectional views?

2. How is the position of a sectional view indicated?

3. How are sectional views identified and labeled?

4. What two purposes do section lines serve in sectional views?

5. What is the main purpose of using auxiliary views?

6. What are the two principal types of auxiliary views?

7. How are auxiliary views identified?

8. Why are detail views sometimes used on blueprints?

9. What are the two main forms of detail views?

10. What type of sectional view is shown in SECTION AA of Figure 5-26?

11. What type of section view is shown at Ⓐ?

12. What type of sectional view is shown at Ⓑ?

13. Identify the materials shown by lines A, B, E, F.

14. Identify lines C, D, G, H.

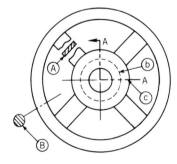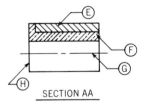

Figure 5-26

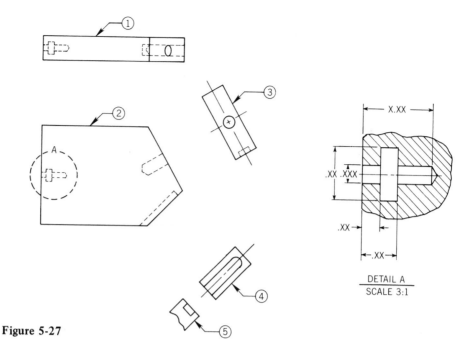

Figure 5-27

DETAIL A
SCALE 3:1

15. Identify views 1-5 in Figure 5-27:

16. What type of detail view is shown at **A** ?

17. What is the scale of the detail view?

18. What is the difference between the two types of auxiliary views?

Answers To Self Test

1. To show clearly the shape and form of internal details

2. By a cutting plane line

3. By letters that refer to the specific cutting plane

4. Show the exposed edges of sectional views; identify the materials used to make the part

5. Show the true shape, size and form of inclined surfaces

6. Primary and secondary auxiliary views

7. By the principle view with which they are aligned.

8. To show details in a form that is easier to see than that shown in the principle view.

9. Enlarged detail and sectional detail views

10. Half sectional view

11. Revolved sectional view

12. Removed sectional view

13. A = Cast iron

B = Cast iron

E = Cast iron

F = Brass, bronze, or other copper alloy

14. C = Cutting plane line

D = Hidden line

G = Center line

H = Object line

15. 1 = Top view

2 = Front view

3 = Front auxiliary view (primary)

4 = Front auxiliary view (primary)

5 = Secondary auxiliary view

16. Sectional detail view

17. 3 = 1, or drawn object is three times the size of the actual part

18. Primary auxiliary views are aligned with a principal view whereas secondary auxiliary views are aligned to primary auxiliary views.

Fundamentals of Dimensioning

Dimensions are the numerical values that describe, or show, the size or location of features in a blueprint. To interpret a blueprint properly and to make any part accurately, you must know how to read print dimensions.

TYPES OF DIMENSIONS

To describe the size or location of features completely, several different types of dimensions are commonly used. The five basic types of dimensions are linear, angular, radial, tabular, and coordinate.

LINEAR DIMENSIONS

Linear dimensions (Figure 6-1) show the straight line distance between two points. As you can see, these dimensions are drawn parallel to the surface they describe.

The dimensional units commonly used with linear dimensions are the inch and the millimeter. Inch-based dimensions are expressed on a print either as decimal inches or as fractional inches (Figure 6-2). Decimal inches are more commonly used than fractional inches. Fractional inches generally show only thread sizes or rough stock sizes. Prints drawn to meet the requirements of the SI (Systeme International d'Unites), or International System of units, are dimensioned in millimeters. Millimeters are shown only as decimal values (Figure 6-3).

Figure 6-1 Linear dimensions.

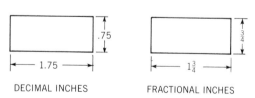

Figure 6-2 Inch-based dimensions.

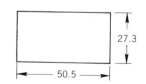

Figure 6-3 Millimeter-based dimensions.

ANGULAR DIMENSIONS

Angular dimensions show the size of angular details. Depending on their application, angular dimensions appear as either linear or angular values.

Figure 6-4 The length of this arc in inches is 3.75 in. In millimeters, its length is 93.4 mm.

Figure 6-5 Placement of dimensions for arcs and chords.

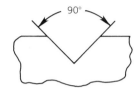

Figure 6-6 Dimensioning angles with angular units.

Dimensions that show the length of an arc or chord are shown in either inch or millimeter units (Figure 6-4). An arc is a part of a curved line, or circle, between two points. A chord is a straight line that connects the ends of an arc. This difference is shown in Figure 6-5. Be careful not to confuse these dimensions. Their values are quite different.

Another difference between these dimensions is the shape of the dimension lines. Dimension lines that show the length of an arc are curved; those that show the length of a chord are straight.

Dimensions that show the size of an angle are expressed in angular units. These units are degrees, minutes, and seconds. Although not all angular dimensions will contain all three values, you should become familiar with each and how they relate to the total measurement. Each circle contains 360° (° = degrees); each degree contains 60′ (′ = minutes); and each minute contains 60″ (″ = seconds). In addition to using angular units, these dimensions also use a curved dimension line as shown in Figure 6-6.

RADIAL DIMENSIONS
Radial dimensions indicate the size of a radius. As shown in Figure 6-7, some of the variations of radial dimensions are radii dimensions, spherical radii dimensions, shortened radii dimensions, and true radii dimensions.

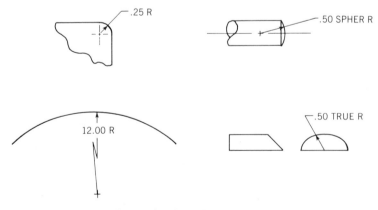

Figure 6-7 Radial dimensions.

Radii dimensions appear with leader lines. Leader lines show the size of radii such as rounded corners or similar details. *Spherical radii* appear with the note "SPHER R" and indicate that the radii shown in the views are spherical. *Shortened radii* are seen when a large radius must be shown in a limited space or where the center position of the radius would appear off the drawing sheet. Shortened radii are indicated by a leader with a zigzag. *True radii* are shown with the note "TRUE R." The note indicates that the radius is true regardless of how it may appear in the blueprint.

TABULAR DIMENSIONS

Tabular dimensions show the sizes of repeated details, such as the holes in Figure 6-8. Using tabular dimensions for parts like this prevents the print from being cluttered with an excessive number of dimensions and helps to reduce the chance of misinterpretation.

Another common use of tabular dimensions is to show the sizes of parts with identical forms but different sizes (Figure 6-9). Here the basic shape of the part is the same even though the sizes change with each particular part number.

In either case, tabular dimensions are shown in a chart, and the dimensions they refer to are indicated on the print with letter symbols.

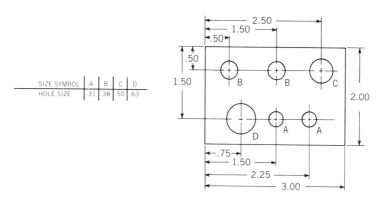

Figure 6-8 Tabular dimensions used to eliminate clutter.

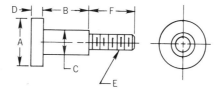

PART NUMBER	DIMENSIONS					
	A	B	C	D	E	F
123-1	.75	.75	.38	.25	$\frac{1}{4}$-20-UNC	.75
123-2	1.00	1.00	.50	.25	$\frac{5}{16}$-18-UNC	1.00
123-3	1.25	1.25	.75	.38	$\frac{3}{8}$-16-UNC	1.25
123-4	1.50	1.50	1.00	.38	$\frac{1}{2}$-13-UNC	1.50

Figure 6-9 Tabular dimensions used to show different sizes of parts with the same shape.

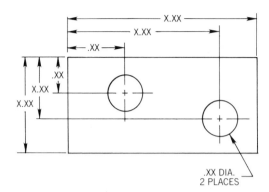

Figure 6-10 Coordinate dimensions.

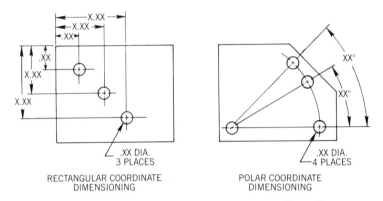

RECTANGULAR COORDINATE
DIMENSIONING

POLAR COORDINATE
DIMENSIONING

Figure 6-11 Rectangular and polar coordinate dimensions.

COORDINATE DIMENSIONS

Coordinate dimensions show the sizes of features as they relate to one reference point (Figure 6-10). These dimensions are frequently referenced from an edge, corner, or center axis, as shown. One of the most common applications of coordinate dimensioning involves numerical control equipment. The two major forms of coordinate dimensions are rectangular and polar (Figure 6-11).

**BASIC RULES
FOR READING
DIMENSIONS**

To interpret a print properly, you should learn the basic rules of reading dimensions.

1. *Never Measure a Print.* Drafters are responsible only for the accuracy of the dimensions. The drawn object does not have to be drawn to an exact scale. For this reason, there may be some difference between the drawn object and the actual dimensioned size. When there is no dimension shown, look at the other views in the print to see if you can calculate the missing dimension. If you cannot find or calculate the required dimension, show the print to your supervisor. *Never measure a print to find a missing dimension.*

2. The scale of a blueprint has no effect on the sizes shown in the print. The dimensions given in a blueprint always refer to the final, finished size of the part. The scale indicates only that the drawn object is proportional to the actual object.

3. All dimensions on a print are considered to be either working dimensions or reference dimensions. *Working dimensions* are those that control the size of the part in the print. *Reference dimensions* are put on a print only for convenience. These dimensions give general information rather than specific instructions. Reference dimensions are shown on a print with parenthesis [()] or with the abbreviation REF (Figure 6-12). Another dimension you may find is the NOT TO SCALE dimension. These dimensions are indicated by a straight line under the dimension. NOT TO SCALE dimensions are used in cases where a dimension is changed and the drawn detail it refers to is left unchanged. This may make the size of the drawn object appear incorrect when compared to its dimension (Figure 6-13).

4. Dimensions are normally placed in the view that best describes the detail. Thus dimensions are not positioned to show the size or location of a detail from a hidden line. Details such as holes, radii, and angles are dimensioned in the view that show their true form.

5. Dimensions are generally shown as two-place decimals unless additional precision is required. Then either three- or four-place decimals are used (Figure 6-14).

6. For dimensional values less than 1 inch, zeros *are not* used before the decimal point. However, for dimensional values less than one millimeter, zeros *are* used before the decimal point (Figure 6-15).

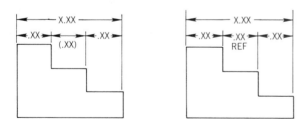

Figure 6-12 Reference dimensions.

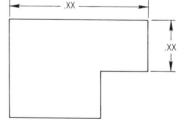

Figure 6-13 Not to scale dimensions.

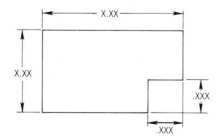

Figure 6-14 Two-place decimals are used for most dimensions.

INCH DIMENSIONS

.50 NOT 0.50
.25 NOT 0.25

MILLIMETER DIMENSIONS

0.50 NOT .50
0.25 NOT .25

Figure 6-15 Proper use of zeros for inch and millimeter dimensions.

7. Except where common hardware or threads are dimensioned, fractional and decimal sizes should not be mixed on a print.

8. Dimensional values greater than 1000 *do not* take a comma (,) in the dimension.

9. Dimensional values shown in a print do not have the dimensional units (in. or mm) indicated. Each print should have the note:

UNLESS OTHERWISE SPECIFIED ALL DIMENSIONS ARE IN INCHES

or

UNLESS OTHERWISE SPECIFIED ALL DIMENSIONS
ARE IN MILLIMETERS

TOLERANCE AND ALLOWANCE

In any machining operation, the parts must conform to the size shown in the print. In most cases, however, it is impossible to make every part to the exact size. For this reason, designers and drafters specify a range of sizes for each dimension. This range is called the *tolerance*. When two or more parts must be assembled, the difference in the sizes that allows one member to fit the other is called the *allowance*. To get the most efficiency for the least possible cost, you must pay close attention to the tolerance and allowance values. The first step in interpreting tolerances and allowances is to learn the terms commonly used to describe these values.

NOMINAL SIZE

The nominal size is used for general identification. For example, a steel rod may be identified as .500″ diameter. But if the rod were measured, it might be found to vary between .497″ to .503″. The nominal size is intended only to give a general identification, not an exact size. Another example of the nominal size is in lumber. Although a board may be called a 1″ × 2″, the actual measured size of the board is only $\frac{3}{4}$″ × $1\frac{1}{2}$″. Here the nominal size is 1″ × 2″.

ACTUAL SIZE

The actual size is the measured size of the dimension. An actual size of .500″ will measure .500″.

BASIC SIZE

The basic size is the basic design size of the part. In practice, the basic size is the size to which the tolerance is applied.

TOLERANCE

The tolerance is the total allowable variation from the basic size of a part. Tolerance is generally specified in either of two ways: *unilateral* or *bilateral*.

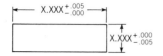

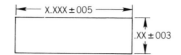

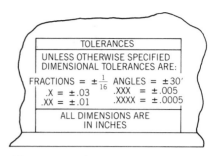

Figure 6-16 Unilaterial tolerance.   **Figure 6-17** Bilaterial tolerance.

Figure 6-18 General tolerance note.

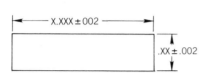

Figure 6-19 Specific tolerance values.

Unilateral tolerance is a tolerance that allows variation in only one direction (Figure 6-16). Bilateral tolerance is a tolerance that allows variation in either a plus or a minus direction (Figure 6-17).

Tolerance may be specified either as a *general tolerance* or as a *specific tolerance*. General tolerances are given in a tolerance note in the print (Figure 6-18). When a dimension on a print does not have a tolerance applied directly, the value shown in the general tolerance note should be used. When reading a general tolerance note, the values refer to the decimal places of the dimension. For example, if a general note read: .XX = ±.01, it would mean all two-place decimal dimensions have a tolerance of ±.01. Specific tolerances are added to the basic dimension shown on the print (Figure 6-19). When specific tolerances are used, they are always used in place of the general tolerance values. Specific tolerances indicate a special situation that cannot be covered by the general tolerance. For example, if a hole size were critical to the operation of the part, a closer specific tolerance would be used. Likewise, if the general tolerance stated a tolerance closer than that required by the part, a specific tolerance may be used to allow a larger tolerance.

Always pay close attention to the tolerance value. Making a part too accurately is almost as wasteful as making one out of tolerance. The values shown mean any part made within the range of sizes specified will work. Trying to make every part exactly to the basic size is very time consuming and adds nothing to the value of the part.

LIMITS

Limits are the maximum and minimum sizes permitted by the tolerance. For example, as shown in Figure 6-20, the dimension is .500 ± .005. The limits of this dimension are .505″ and .495″.

Figure 6-20 Limits of size.

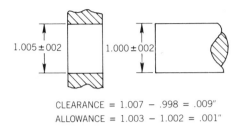

CLEARANCE = 1.007 − .998 = .009"
ALLOWANCE = 1.003 − 1.002 = .001"

Figure 6-21 Calculating clearance and allowance.

ALLOWANCE

Allowance is the allowed difference in sizes of mating parts that permits the desired fit. In Figure 6-21, the shaft size is 1.000 ± .002 and the hole size is 1.005 ± .002. By comparing the limits of these two parts, you can easily find the clearance and the allowance. The largest hole size is 1.007". The smallest shaft size is .998". By subtracting .998" from 1.007", the clearance is found to be .009". To calculate the allowance, simply subtract the largest shaft size from the smallest hole size. Or, 1.003 − 1.002 = .001". The allowance is .001". When the hole size is larger than the shaft size, the assembly is said to be a *clearance fit*. In cases where the shaft is larger than the hole, an *interference fit* results.

Remember that the allowance is the minimum allowable difference between mating parts. The clearance is the maximum allowable difference between mating parts. In formula form, this becomes:

Allowance = Smallest hole size − largest shaft size

Clearance = Largest hole size − smallest shaft size

BASIC RULES FOR TOLERANCES

Tolerances, like dimensions, are applied to a print by following certain rules. To interpret these values properly, you should know and apply these rules.

1. Tolerance values should never be assumed by the number of decimal places in the basic dimension. Always check the general tolerance note if there is no tolerance on the dimension.

2. Dimensions and tolerance values must have the same number of decimal places for inch-based dimensions (Figure 6-22).

3. When millimeter dimensions have a unilateral tolerance and the plus or minus value is zero, a single zero may be shown without a plus or minus sign (Figure 6-23).

4. When millimeter dimensions have a bilateral tolerance, both the plus and minus values must have the same number of digits (Figure 6-24).

5. Limit dimensions must have the same number of decimal places in both inch and millimeter dimensioning. Limit dimensions state the tolerance in the dimension rather than in the plus or minus method. Limit dimensions are shown in Figure 6-25.

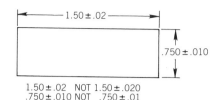

$$27 {+0.03 \atop 0} \quad \text{OR} \quad 27 {0 \atop -0.03}$$

Figure 6-23 Zeros used in unilateral tolerances for millimeter-based dimensions.

1.50±.02 NOT 1.50±.020
.750±.010 NOT .750±.01

Figure 6-22 Dimensional and tolerance value must have the same number of decimal places for inch-based dimensions.

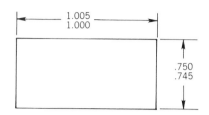

$$37 {+0.25 \atop -0.10} \quad \text{NOT} \quad 37 {+0.25 \atop -0.1}$$

Figure 6-24 Bilateral tolerance values must have the same number of decimal places in millimeter-based dimensions.

Figure 6-25 Limit dimensioning.

DIMENSIONAL PLACEMENT

Dimensions are placed on a print in one of two ways, either *aligned* or *undirectional.* In the aligned method, the dimensions are placed parallel to the surface they control (Figure 6-26). Unidirectional dimensions are placed on the drawing so each dimension can be read parallel to the bottom of the drawing sheet (Figure 6-27).

Another method, or variation, of dimensioning is the *staggered method.* Staggered dimensions (Figure 6-28) are used when several dimensions are located close together. Rather than crowding the dimensions together, the dimension lines are broken at different places to make the dimensions easier to read.

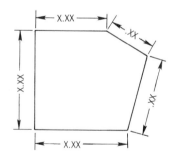

Figure 6-26 Aligned dimensions.

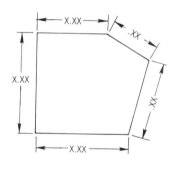

Figure 6-27 Unidirectional dimensions.

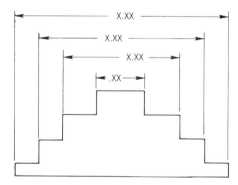

Figure 6-28 Staggered dimensions.

DUAL DIMENSIONING

As the use of the SI continues to increase throughout industry, more and more prints are being dual dimensioned. *Dual dimensioned prints* are blueprints that have the dimensional values expressed in both inch and millimeter units. The two most common ways to show both dimensional units on a single print are *direct dual dimensioning* and *conversion chart dimensioning.*

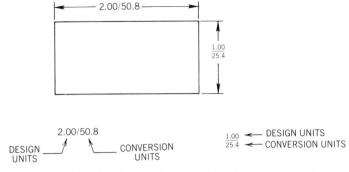

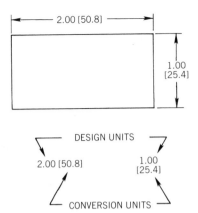

Figure 6-29 Position method of direct dual dimensioning.

Figure 6-30 Bracket method of direct dual dimensioning.

DIRECT DUAL DIMENSIONING

Direct dual dimensioning uses both the inch and millimeter values for each dimension. Two methods of direct dual dimensioning are the *position method* and the *bracket method.* In the position method, the design units are shown above a horizontal line or to the left of a slash (Figure 6-29). *Design units* are the dimensional units that the part was originally designed to suit. These values tend to be more rounded than the conversion values. The conversion values are located below the horizontal line or to the right of the slash, as shown. In the bracket method, the design units are shown as normal dimensions and the conversion units are shown in brackets (Figure 6-30).

The main advantage of direct dual dimensioning is the direct conversion on each dimension. This permits quick cross reference between inch and millimeter dimensions. The principal disadvantages to using direct dual dimensioning are the added chance of misreading a dimension and the additional clutter to the print.

CONVERSION CHART DIMENSIONING

The conversion chart method of dual dimensioning uses the design units to dimension the part. The equivalent conversion values are shown in a chart on

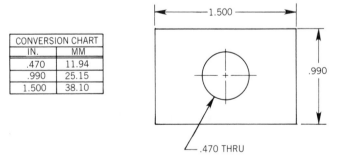

CONVERSION CHART	
IN.	MM
.470	11.94
.990	25.15
1.500	38.10

Figure 6-31 Conversion chart method of dual dimensioning.

the print. This method eliminates the clutter common to direct dual dimensioning and is also easier to read. Only the dimensions shown on the print and their equivalents are shown in the chart. This eliminates unnecessary conversions and simplifies reading the chart. The chart itself is normally made with the design units in the left column and the converted values listed to the right. The dimensions are generally listed from smallest to largest (Figure 6-31).

SELF TEST

1. List the five basic types of dimensions.

2. What two dimensional units are commonly used to show dimensions of length with inch-based dimensioning?

3. What dimensional unit is commonly used to show dimensions of length with millimeter-based dimensioning?

4. What three dimensional units are used to show angular dimensional values? Also show the abbreviation used to denote each value.

5. Identify dimensions a, b, c in Figure 6-32.

6. Identify dimensional forms a-f in Figure 6-33.

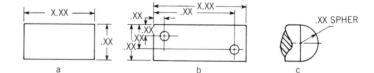

Figure 6-32   a   b   c

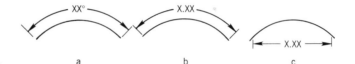

Figure 6-33   a   b   c

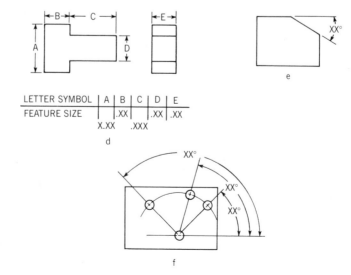

7. Why should you *never* measure a print?

8. What effect does the drawing scale have on the print dimensions?

9. What type of dimension is shown with a line under the dimension?

10. What type of dimension is shown in parentheses?

11. What is a nominal size?

12. What is a basic size?

13. Define tolerance.

14. Define allowance.

15. Define limits.

16. What two types of tolerances are commonly used to control part sizes?

17. What is the allowance in the assembly shown in Figure 6-34?

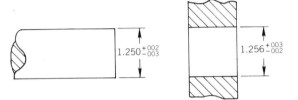

1.250^{+002}_{-003} 1.256^{+003}_{-002}

Figure 6-34

18. What is the clearance in this assembly?

19. What type of fit will result from the part sizes shown in this assembly?

20. What three types of dual dimensioning are commonly used on prints?

21. What three views are used to describe the part in Figure 6-35?

22. Identify lines G-L.

23. Find dimensions A-F.

24. What size radius is used on the corners?

25. How many radii are called out?

26. What type of dimension shows the radii?

27. What angle is shown?

28. What type of dimension is used to show the size of the angle?

29. What tolerance is applied to the angle?

30. What tolerance is applied to the other print dimensions?

31. What is the outside diameter of the part in Figure 6-36?

32. What is the hub diameter?

33. Which of these two diameters has the closest tolerance?

34. What type of dimension is used to show the sizes of the holes in the flange?

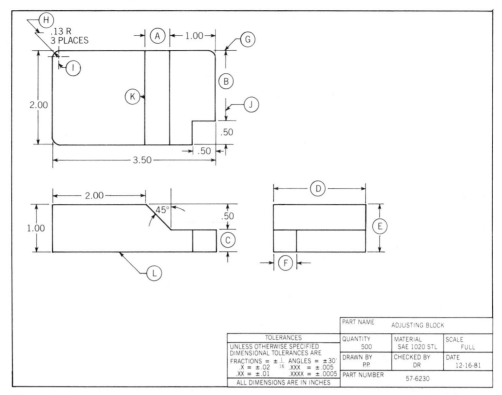

Figure 6-35

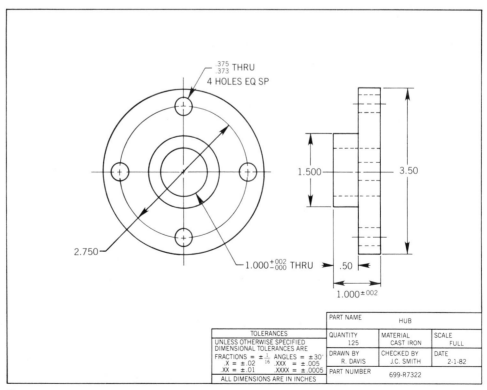

Figure 6-36

35. What is the diameter of the bolt circle?

36. What type of tolerance is used to show the tolerance of the center hole?

37. What type of tolerance is used to show the tolerance of the part thickness?

38. What is the tolerance of the bolt circle?

39. What are the limits of size of the center hole?

40. What are the limits of size of the part thickness?

**Answers To
Self Test**

1. a. Linear dimensions

    b. Angular dimensions

    c. Radial dimensions

    d. Tabular dimensions

    e. Coordinate dimensions

2. a. Decimal inches

    b. Fractional inches

3. Millimeter

4. a. Degree (°)

    b. Minute (')

    c. Second (")

5. a. Size of angle

    b. Length of arc

    c. Length of chord

6. a. Linear dimensions

    b. Rectangular coordinate dimensions

    c. Spherical radii, radial dimension

    d. Tabular dimensions

    e. Angular dimension

    f. Polar coordinate dimension

7. Because the drafter is not responsible for the scale of the drawn view, only the dimension. Drawn views are often not the size they should be.

8. None at all. The stated size is that of the finished part.

9. NOT TO SCALE dimensions

10. Reference dimension

11. The size used for general identification of the part

12. The size to which the tolerance value is applied

13. The total allowable variation in size from the basic size

14. The allowed difference in sizes between mating parts that permits the desired fit

15. The maximum and minimum sizes a part tolerance will permit.

16. Specific tolerances and general tolerances.

17. .002″

18. .012″

19. Clearance fit

20. a. Direct dual dimensioning, position method

    b. Direct dual dimensioning, bracket method

    c. Conversion chart dimensioning

21. Top, front, and right side

22. G = Extension line

    H = Leader line

    I = Center line

    J = Dimension line

    K = Object line

    L = Object line

23. A = .50″

    B = 1.50″

    C = .50″

    D = 2.00″

    E = 1.00″

    F = .50″

24. .13 R

25. 3

26. Radial dimension

27. 45°

28. Angular dimension

29. ± 30′

30. ± .01″

31. 3.50″

32. 1.500″

33. Hub diameter

34. Limit dimensions

35. 2.750″

36. Unilateral

37. Bilateral

38. ± .005″

39. 1.002 and 1.000

40. 1.002 and .998

UNIT 7   # Dimensioning Machined and Fabricated Details

Dimensions show the specific size, shape, and location of part features. Until now we have discussed only those dimensions that relate to the overall size of a part. In addition to these dimensions, however, other dimensions are also used to show the size or characteristics of machined or fabricated details. Machined and fabricated details are normally shown with a combination of standard dimensions and dimensional notes. To interpret these properly, you must understand both the dimension and the methods used to make the detail. This unit covers the dimensional notes, or *callouts*, most often used on blueprints.

DIMENSIONING HOLES

The most common machined detail found on any part is the hole. Holes are generally easy to interpret on a print. But to make sure you can read these dimensions properly, we will cover the two basic forms of holes common to manufactured parts: the *drilled hole* and the *slotted hole.*

DRILLED HOLES

Drilled holes are normally specified on a print by their diameter and depth (Figure 7-1). When the hole goes completely through the part, the note THRU will appear in place of the depth. These dimensions do not specify the operation used to produce the holes. This decision is left to the shop. Only in rare cases, when engineering control is desired, will a method be specified for producing these holes.

Many times, drilled holes will be grouped together in a hole pattern. When these patterns are dimensioned, the hole spacing will be added to the diameter and depth in the dimensional callout (Figure 7-2). Circular hole patterns may specify the spacing as EQUALLY SPACED or in degree increments (Figure 7-3). In either case, the center line that shows the center of the circular pattern is called the *bolt circle* and is always included, and dimensioned, on the print.

SLOTTED HOLES

Slotted holes are a variation of the standard hole used, in most cases, to provide adjustment in an assembly. Either a dimensional note or a drawn dimension

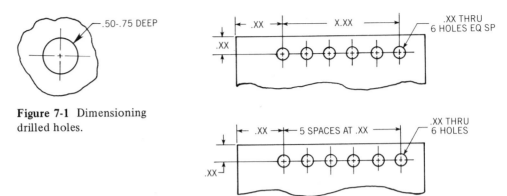

Figure 7-1 Dimensioning drilled holes.

Figure 7-2 Dimensioning straight hole patterns.

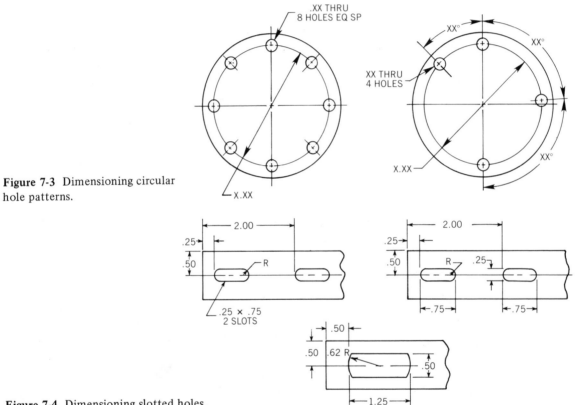

Figure 7-3 Dimensioning circular hole patterns.

Figure 7-4 Dimensioning slotted holes.

shows the overall size of the slot. When the radius of the ends equals half the width of the slot, the note "R" on a leader is used. When the end radius is a size other than half that of the slot width, the radius is specified, as shown in Figure 7-4.

DIMENSIONING SECONDARY OPERATIONS IN HOLES

Secondary operations are processes that improve a hole's appearance, size, finish, accuracy, or that modify the hole for a certain purpose. The most common secondary operations are reaming, boring, undercutting, countersinking, chamfering, counterboring, and spotfacing.

REAMING

Reaming enlarges drilled holes to an exact size. Reamed holes are indicated on a print either by a direct note or by a close tolerance on the hole size (Figure 7-5). As a rule, reamed holes are straighter, smoother, and closer to the required size than are drilled holes.

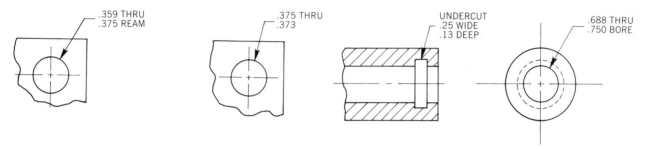

Figure 7-5 Dimensioning reamed holes.

Figure 7-6 Dimensioning bored and undercut holes.

BORING AND UNDERCUTTING

Boring is the process of enlarging drilled holes with a single-point boring tool. Like reaming, boring also straightens and smoothes holes, but boring is not limited to specific sizes. Undercutting is a variation of the boring operation for machining grooves inside holes. These details are specified on a print as shown in Figure 7-6.

COUNTERSINKING AND CHAMFERING

Countersinking and chamfering produce conical surfaces on the edge of a hole. Countersinking is primarily used to produce a seat for flathead screws (Figure 7-7). Chamfering is a similar process, but it only removes the burred edges of a hole (Figure 7-8). When these operations are specified on a print, sizes may be shown in two ways. Countersunk holes appear with a note specifying the diameter at the top of the countersunk area and the angle of the countersink (Figure 7-9). The standard angles used for countersinking are 82°, 90°, and 100°. Chamfered holes are specified by the depth of the chamfer and the angle of one side (Figure 7-10). The total, or included angle of a normal chamfer is 90°.

Figure 7-8 Chamfers remove burred edges.

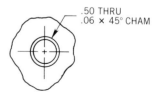

Figure 7-10 Dimensioning a chamfered hole.

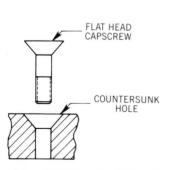

Figure 7-7 Countersinking for flathead screws.

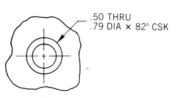

Figure 7-9 Dimensioning a countersunk hole.

Figure 7-11 Counterbores and spotfaces.

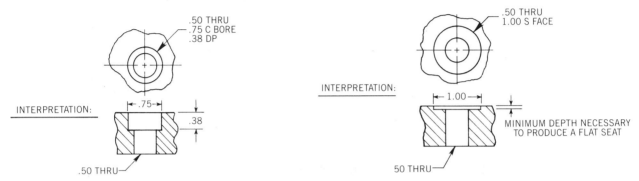

Figure 7-12 Dimensioning counterbores.

Figure 7-13 Dimensioning spotfaces.

COUNTERBORING AND SPOTFACING

Counterboring and spotfacing produce an enlarged, flat bottom area around the top of a hole. This enlarged area becomes a flat seat for bolts and nuts. The main difference between counterboring and spotfacing is the depth of the enlarged area. Counterbored holes are made deeper to set the head of the bolt below the surface of the part. Spotfaced holes are machined only deep enough to provide a flat seat for the bolt head (Figure 7-11). Counterbored holes are specified on a print by their diameter and depth (Figure 7-12). Spotfaced holes are normally specified by their diameter only; the depth is left to the machine operator (Figure 7-13).

DIMENSIONING EXTERNAL MACHINED DETAILS

External machined details are those machining processes performed on the outside surfaces of a part. The most common types of external machined details are grooves, chamfers, fillets, rounds, radii, knurls, and flats.

GROOVES

External grooves are used for "O" ring seats, bearing seats, and for thread runout. These details are dimensioned several ways, depending on the design or application. It is very important to pay close attention to exactly where the dimension is placed and what the dimension specifies. When the depth is critical, the part may be dimensioned from the outside edge. When the diameter at the bottom of the groove is most important, this diameter may be specified (Figure 7-14). When external grooves have a radius at the bottom, the groove is normally dimensioned with the radius size rather than with a width dimension.

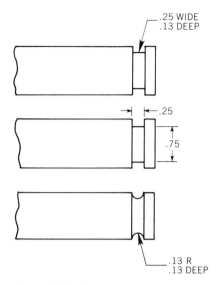

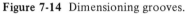

Figure 7-14 Dimensioning grooves.

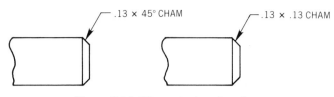

Figure 7-15 Dimensioning chamfers.

Figure 7-16 Rounds and fillets.

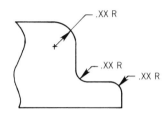

Figure 7-17 Dimensioning rounds, fillets, and radii.

CHAMFERS

External chamfers, like internal chamfers, remove the burred edges of a part. These details are dimensioned in the same way as are internal chamfers—by their length and angle. Another way to show the size of a chamfered edge is by dimensioning the length of both sides of the chamfer, as shown in Figure 7-15.

FILLETS, ROUNDS, AND RADII

Fillets and rounds are normally found on cast or forged parts. These terms identify the rounded areas on the internal and external edges and corners of these parts (Figure 7-16). Radii are the rounded corners and edges on machined parts. When these details are dimensioned, the value shown is the radius of the rounded form (Figure 7-17).

KNURLS

Knurling is the rolling of an impression into the external surface of a shaft. The two main kinds of knurling used in most shops are *straight knurling* and *diamond knurling* (Figure 7-18). Straight knurling generally provides a raised surface for press fits. Diamond knurling is normally used for decorative hand grips on tools. When knurling is specified on a print, the application determines how it is dimensioned. Decorative knurls are normally dimensioned in a note showing the length and pitch of the knurl. Press fit knurls are also dimensioned

STRAIGHT   DIAMOND

Figure 7-18 Straight and diamond knurls.

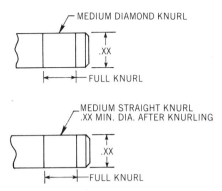

Figure 7-19 Dimensioning knurls.

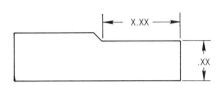

Figure 7-20 Dimensioning flats.

by length and pitch of the knurl, but when the press fit size is critical, the outside diameter of the part after knurling is also shown. Typical methods of dimensioning knurls are shown in Figure 7-19.

FLATS

Flats are machined details that provide a flat bottom seat on shafts for setscrews or similar fasteners. When these details are specified, the flat is dimensioned by the length of the flat and the distance from the flat to the opposite side of the shaft (Figure 7-20).

DIMENSIONING THREADED PARTS

From common hardware to complex threaded parts, the screw thread plays a major role in mechanical assemblies. To identify these details properly on a print, you should know the common thread variations, how threads are shown in prints, and the standard methods used to dimension threads.

SCREW THREADS

Throughout manufacturing there are many different kinds of screw threads. However, the four most common types are the *unified thread, metric thread, acme thread,* and the *pipe thread.*

1. *Unified threads* are the most common threads used for general purpose threaded parts. The basic form of this thread has a 60° thread angle with a rounded root (Figure 7-21).

2. *Metric threads* are similar to unified threads in that both have a 60° thread angle. But the pitch, or spacing of the threads, is different, and the two threads are not interchangeable. The basic metric thread is shown in Figure 7-22.

3. *Acme threads* are a variation of the screw thread commonly used for transmission of motion, power and torque. The basic acme thread has a 29° thread angle (Figure 7-23).

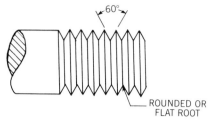

Figure 7-21 Unified thread form.

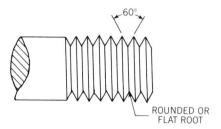

Figure 7-22 Metric thread form.

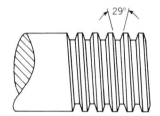

Figure 7-23 Acme thread form.

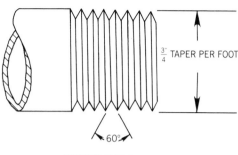

TAPERED PIPE THREAD

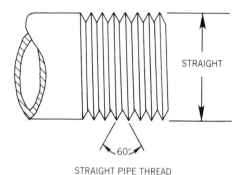

STRAIGHT PIPE THREAD

Figure 7-24 Pipe thread form.

4. *Pipe threads* are commonly found in two forms, tapered and straight. *Tapered pipe threads* are normally used for self-sealing joints such as those in water pipes or air lines. The basic tapered pipe thread is a 60° thread angle with a taper, or conical form, of $\frac{3}{4}''$ per foot. *Straight pipe threads* are identical to tapered pipe threads except that they have no taper. Straight pipe threads are often used when a self-sealing joint is not required, such as in electrical conduit. The basic shape of these threads is shown in Figure 7-24.

SCREW THREAD TERMS

Screw threads, like other machined details, have their own terminology describing various parts of the thread. To interpret thread designations properly, you should learn these terms and how each relates to the thread itself.

As shown in Figure 7-25, the *nominal diameter* is the outside diameter of the thread. This is also called the *major diameter*. The *pitch diameter* is the diameter of an imaginary circle approximately half the distance between the root and crest of the thread. This is the diameter along which properly cut threads

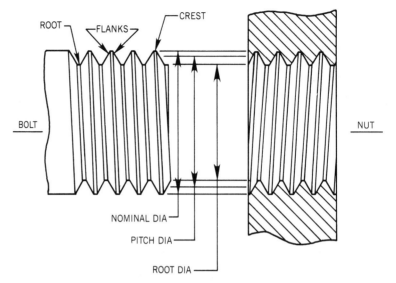

Figure 7-25 Parts of a screw thread.

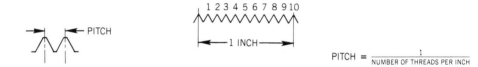

Figure 7-26 Pitch of a thread.   **Figure 7-27** Calculating the pitch.

will engage. The *root diameter* is the diameter of the fastener at the bottom of the threads. The top surface of each thread is called the *crest*. The bottom point is called the *root*. The sides of the thread are called the *flanks*. The *pitch* of a thread is the distance between adjacent threads, measured between identical points on the adjacent threads (Figure 7-26).

The pitch of a thread is an important value. You should understand its meaning completely. On inch threads, the pitch is normally expressed as the number of threads per inch. However, this is not totally correct. The actual value of the pitch is equal to 1 divided by the number of threads per inch (Figure 7-27). So a thread that has 10 threads per inch has a pitch of $1 \div 10$, or $.100''$. All metric threads are identified by their pitch, which is read as part of the thread designation.

READING THREAD DESIGNATIONS

Thread designations are a standard system of numbers and letters that describe and identify each value in a screw thread. To make any threaded part properly, you must be able to read and understand the designations for each type of screw thread.

1. *Unified threads* are shown in a print by the designation indicated in Figure 7-28. As you can see, each element in this designation contains a single

piece of information necessary to make the thread. The first number, $\frac{3}{4}$, shows the nominal diameter of the thread. The second number, 10, indicates the number of threads per inch. The next series of letters, UNC, shows the thread series. The last number and letter combination, 2A, indicates the class of fit and type of thread. The numbers that show the class of fit are 1, 2, and 3, with 1 the loosest and 3 the tightest fit. The letters A and B indicate whether the thread is external (A) or internal (B). The letters LH, when used, mean the thread is left hand.

Figure 7-28 Unified thread designations.

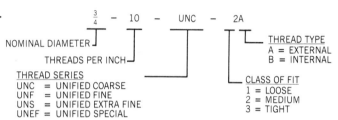

2. *Metric threads* use a designation similar to that for unified threads. The main difference is the added values shown in the metric designation (Figure 7-29). The first letter and number combination, M16, shows the thread is metric (M) and indicates the nominal diameter (16 millimeters). The next digit, 2, is the pitch of the thread in millimeters. The next number and letter combination, 4h, indicates the class of fit of the thread. The final number and letter combination, 5H, indicates the class of fit of the mating thread.

Figure 7-29 Metric thread designations.

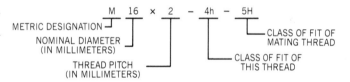

3. *Acme threads* also use a similar numbering system. The main difference is that the word ACME is included in the designation (Figure 7-30). The first number in an acme designation, $\frac{3}{4}$, is the nominal diameter of the thread. The second number, 6, indicates the number of threads per inch. The word ACME identifies the thread type. And the last number and letter combination, 2G, specifies the class of fit and application. The numbers 2, 3, 4, and 5 show the class of fit of acme threads; 5 is the tightest and 2 is the loosest fit. The letter G indicates the thread is general purpose. If a "C" were used, it would indicate that the thread is self-centering.

Figure 7-30 Acme thread designations.

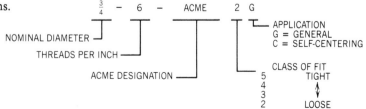

4. *Pipe threads* are also shown with a basic number and letter system (Figure 7-31). The first number in this designation, $\frac{3}{4}$, is the nominal diameter of the thread. The second number, 14, indicates the number of threads per inch. The letters NPT are used to indicate tapered pipe threads. The letters NPS indicate straight pipe threads.

Figure 7-31 Pipe thread designations.

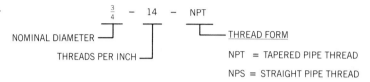

REPRESENTING SCREW THREADS

Three methods are commonly used to show screw threads on a part: pictorial, schematic, and simplified (Figure 7-32). The pictorial is the most complicated and is not often used for shop prints. The schematic and simplified are simple and easy to draw. These are the ones you are most likely to see.

Figure 7-32 Methods used to show threads in a print.

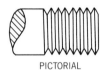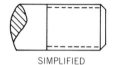

PICTORIAL   SCHEMATIC   SIMPLIFIED

DIMENSIONING TAPERS

Tapers are machined details often found in machine parts. These details provide a self-centering method for locating machine elements. The two primary types of tapers are *self-holding* and *self-releasing*. Self-holding tapers have a taper angle of less than 3°; they will automatically lock into a taper socket of the same form. Self-releasing tapers normally have a taper angle greater than 8°. These tapers must be held in their taper socket with a mechanical fastener such as a bolt.

The standard units for showing tapers are *taper per foot* (TPF) and *taper per inch* (TPI). Both values describe the difference in size between the large end and the small end of the part with relation to the part length. Figure 7-33 shows the standard methods and formulas used to calculate these values.

As shown in Figure 7-34, tapers may be specified in many ways. When a standard taper form is specified, such as #2 MORSE TAPER, the required values for the taper can be found in any machinist's handbook.

DIMENSIONING KEYWAYS

The term keyway is often used to describe a complete keyed assembly. In practice, this assembled unit is made up of three different parts: a *key, keyseat,* and *keyway* (Figure 7-35). Keyed assemblies generally provide a positive drive for

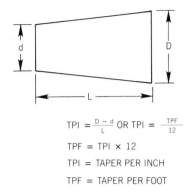

$$TPI = \frac{D-d}{L} \text{ OR } TPI = \frac{TPF}{12}$$

$$TPF = TPI \times 12$$

TPI = TAPER PER INCH

TPF = TAPER PER FOOT

Figure 7-33 Calculating taper values.

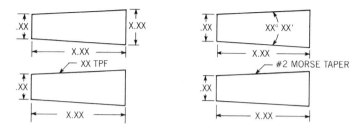

Figure 7-34 Dimensioning tapers.

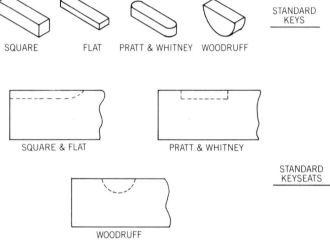

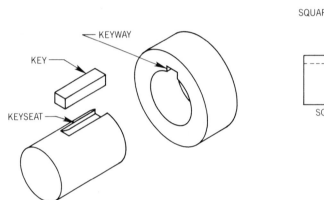

Figure 7-35 Elements of a keyed assembly.

Figure 7-36 Variations of keyed assemblies.

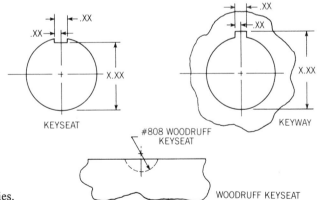

Figure 7-37 Dimensioning keyed assemblies.

units such as pully drives and gears. There are several different kinds of keyed assemblies; the most common are shown in Figure 7-36.

The most common methods of dimensioning keyed assemblies are shown in Figure 7-37. When an assembly is dimensioned with a note, consult an appropriate handbook to find the correct dimensions and tolerance values.

READING DIMENSIONAL NOTES AND SYMBOLS

Dimensional notes and symbols convey specific technical information in as simple a way as possible. As an aid to reading these notes, you must know the standard abbreviations commonly used on prints. A list of the most common abbreviations is given in Table 7-1. You should also be able to read dimensional and process notes quickly and accurately. These notes and symbols are used to

Table 7-1
COMMON BLUEPRINT ABBREVIATIONS

A		Finish all over	FAO	**P**			
Allowance	ALLOW	Flat head	FH	Piece	PC		
Aluminum	ALUM			Pitch	P		
Approved	APPD	**G**		Pitch diameter	PD		
Assemble	ASSM	Gage	GA	Pounds per square inch	PSI		
Assembly	ASSY	Grind	GRD				
Auxiliary	AUX			**R**			
B		**H**		Radius	R		
		Harden	HDN	Ream	RM		
Bevel	BEV	Heat treat	H/T	Reference	REF		
Blueprint	B/P	Hexagon	HEX	Required	REQ		
Bolt circle	BC	High speed	HS	Right hand	RH		
Brass	BRS	High speed steel	HSS	Rockwell	R^c-R^b,		
Brinell	BNL	Horizontal	HORZ		etc.		
Bronze	BRZ	Hour	HR	Round	RD		
Brown & Sharpe	B&S						
Bushing	BUSH	**I**		**S**			
		Inch	IN	Screw	SCR		
C		Included	INCL	Section	SEC		
Cast iron	CI	Information	INFO	Sheet	SH		
Center	CTR	Inside diameter	ID	Specification	SPEC		
Chamfer	CHAM			Spotface	SF		
Circular	CIRC	**K**		Stock	STK		
Clearance	CL	Keyseat	KST	Symbol	SYM		
Condition	COND	Keyway	KWY	Symmetrical	SYMM		
Counterbore	CBORE						
Countersink	CSK	**L**		**T**			
		Left hand	LH				
D		Length	LG	Teeth	T		
Datum	DAT	Linear	LIN	Tensil strength	TS		
Decimal	DEC			Thread	THD		
Degree	DEG	**M**		Thick	THK		
Detail	DET	Magnesium	MAG	Tolerance	TOL		
Diameter	DIA	Major	MAJ	Typical	TYP		
Dimension	DIM	Manual	MAN				
Dowel	DWL	Mark	MK	**V**			
Drawing	DWG	Material	MAT	Variable	VAR		
Drill	DRL	Maximum	MAX	Vertical	VERT		
		Measure	MEAS				
E		Medium	MED	**W**			
Each	EA			Washer	WASH		
Engineering	ENGRG	**N**		Wide	WD		
Equal	EQL	Nominal	NOM				
Estimate	EST	Not to scale	NTS				
		Number	NO				
F		**O**					
Fabricate	FAB	Outside diameter	OD				
Finish	FIN	Overall	OA				

compress a great deal of information into a very limited space. The symbols you are most likely to see on shop prints are surface texture, or finish, symbols, and welding symbols. Although you may not need to know how to interpret welding symbols fully, it is good to know them so you will know how to machine a part that is to be welded.

SURFACE TEXTURE SYMBOLS

The basic surface texture symbol is a check, as shown in Figure 7-38. In addition to the basic symbol, there are a few variations with which you should become familiar (Figure 7-39). These symbols regulate the surface finish of parts that need a controlled finish. In its most common form, the surface texture symbol has a numerical value positioned as shown in Figure 7-40. This value indicates the maximum roughness height in microinches (micrometers). A comparative list of typical values is shown in Table 7-2. In some cases, when control must be very tight, other values are added to the basic symbol. But these values are beyond the scope of this book. Should you need more information, consult the appropriate ANSI standard.

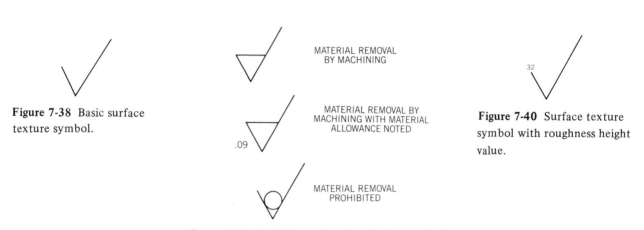

Figure 7-38 Basic surface texture symbol.

MATERIAL REMOVAL BY MACHINING

MATERIAL REMOVAL BY MACHINING WITH MATERIAL ALLOWANCE NOTED

MATERIAL REMOVAL PROHIBITED

Figure 7-40 Surface texture symbol with roughness height value.

Figure 7-39 Variations of surface texture symbols.

Table 7-2
COMPARATIVE ROUGHNESS VALUES

Microinches	Micrometers	Relative Roughness	
1000	25	Very rough	Flame cut
500	12.5	Rough	Sawed
250	6.3	Medium	Forged
125	3.2	Average	Milled, turned, or
63	1.6	Good	similar machining process
32	0.8	Fine	↓
16	0.4	Very fine	Polished
8	0.2	Extremely fine	Lapped or
4	0.1	Extremely fine	superfinished
2	0.05	Extremely fine	↓

WELDING SYMBOLS

Welding symbols are an abbreviated form of instructions weldors must follow to assemble a welded part. Although, as a machinist, you may not be directly concerned with these symbols, it is helpful to know what they mean. The standard welding symbol, and the meaning of each entry, is shown in Figure 7-41.

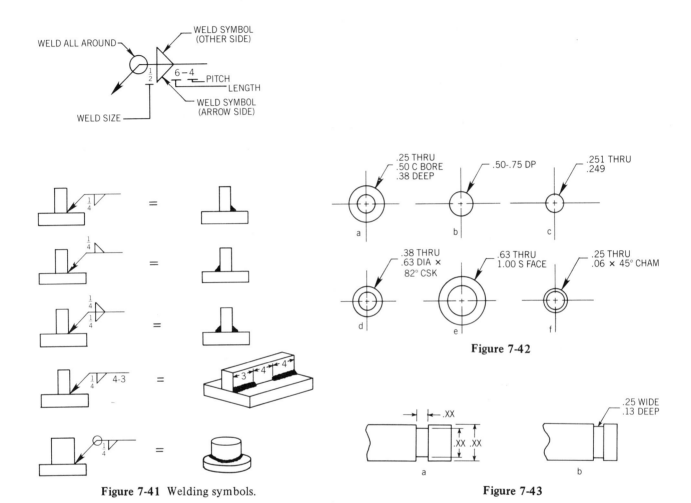

Figure 7-41 Welding symbols.

Figure 7-42

Figure 7-43

SELF TEST

1. Identify the internal machined details a-f in Figure 7-42.

2. Identify the critical sizes of the two grooves shown in Figure 7-43, a and b.

3. What dimension is given for grooves with round bottoms in place of the width?

4. What two types of knurls are commonly used in the shop?

5. What purposes do each of these knurls serve?

6. Identify the values in the unified thread designation shown in Figure 7-44, a-e.

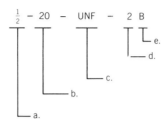

Figure 7-44

7. Identify the values in the metric thread designation shown in Figure 7-45, a-e.

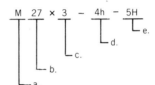

Figure 7-45

8. Identify the values in the acme thread designation shown in Figure 7-46, a-e.

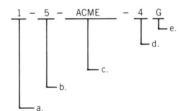

Figure 7-46

9. Identify the values in the pipe thread designation shown in Figure 7-47, a-c.

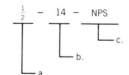

Figure 7-47

10. What two values are used to calculate taper sizes?

11. Identify the three elements in a keyed assembly.

12. What units are used to express the roughness height in a surface texture symbol?

13. What does a circle inside the vee of a surface texture symbol indicate?

14. Find dimensions Ⓐ and Ⓑ in Figure 7-48.

15. What is the taper per inch of the taper on this part?

16. What is the taper per foot of the taper on this part?

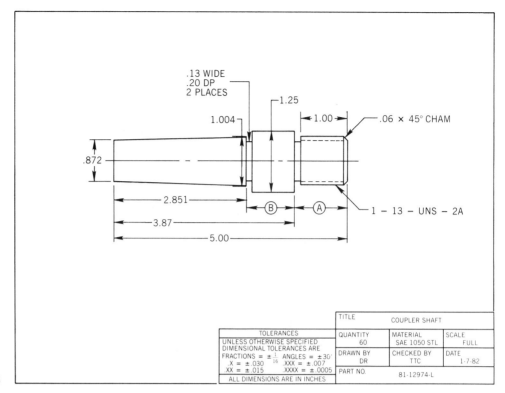

Figure 7-48

17. What is most important in the grooves, the diameter or depth?

18. What does the UNS mean in the thread designation?

19. What is the nominal diameter of the thread?

20. What is the pitch of the thread?

21. What does the letter A indicate in the thread designation?

22. What is the size of the chamfer?

23. Which length dimension has the closest tolerance?

24. What type of thread representation is used to show the threads on this part?

25. What is the tolerance on the angle of the chamfer?

26. What are the limits of size of the center hole in Figure 7-49?

27. What are the limits of size of the .25 diameter hole?

28. What does the symbol √ indicate on the print? What roughness is indicated by this symbol?

29. What is the roughness height of the surface inside the center hole?

30. What type of tolerance is used to tolerance the length of the spacer?

31. How far off center can the .125 dimension be and still pass inspection?

32. What does the Ⓐ indicate?

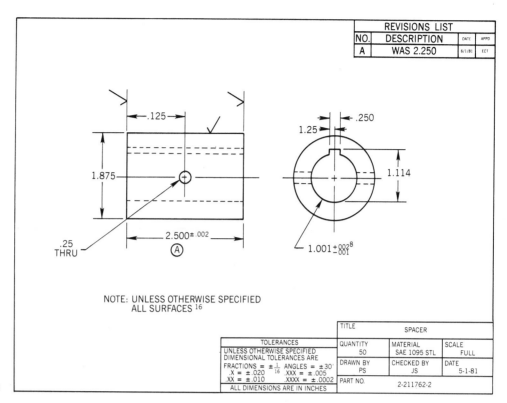

	REVISIONS LIST		
NO.	DESCRIPTION	DATE	APPD
A	WAS 2.250	6/1/81	ECT

NOTE: UNLESS OTHERWISE SPECIFIED
ALL SURFACES 16

TITLE	SPACER		
TOLERANCES UNLESS OTHERWISE SPECIFIED DIMENSIONAL TOLERANCES ARE FRACTIONS = ± 1/16 ANGLES = ±30' .X = ±.020 .XXX = ±.005 .XX = ±.010 .XXXX = ±.0002 ALL DIMENSIONS ARE IN INCHES	QUANTITY 50	MATERIAL SAE 1095 STL	SCALE FULL
	DRAWN BY PS	CHECKED BY JS	DATE 5-1-81
	PART NO.	2-211762-2	

Figure 7-49

33. What is the name of the detail where the key fits in this part?

34. Assuming this part takes a square key, how wide and high should the key be?

35. What is the maximum allowable length of this spacer?

Answers To Self Test

1. a. Drilled and counterbored

   b. Drilled

   c. Drilled and reamed

   d. Drilled and countersunk

   e. Drilled and spotfaced

   f. Drilled and chamfered

2. a. Diameter and width

   b. Depth from outside edge and width

3. Radius of groove

4. Straight and diamond

5. Straight: press fits;

   diamond: decoration

6. a. Nominal diameter

   b. Threads per inch

   c. Thread series: unified fine

   d. Class of fit: medium

   e. Type of thread: internal

7. a. Metric identifier

   b. Nominal diameter

   c. Pitch of thread

   d. Class of fit of this thread

   e. Class of fit of mating thread

8. a. Nominal diameter

   b. Threads per inch

    c. Acme thread

    d. Class of fit

    e. General purpose thread

9. a. Nominal diameter

    b. Threads per inch

    c. Thread series: straight pipe

10. Taper per inch and taper per foot

11. a. key

    b. Keyway

    c. Keyseat

12. Microinches and micrometers

13. Material removal prohibited

14. A = 1.13″

    B = 1.019″

15. .046 TPI

16. .552 TPF

17. Depth from the outside surface

18. Unified special

19. $\frac{3}{4}$″

20. Pitch = $\frac{1}{13}$ = .077″

21. The thread is external

22. .06 long at a 45° angle

23. 2.851″

24. Simplified

25. ± 0° 30′

26. 1.003 and 1,001

27. .26 and .24

28. Surface texture symbol:

    16 microinches

29. 8 microinches

30. Bilateral

31. ± .005″

32. A change as noted in the

    revisions block

33. Keyway

34. .250″ × .250″

35. 2.502″

UNIT 8 # Geometric Dimensioning and Tolerancing

Geometric dimensioning and tolerancing is a simplified method of dimensioning and tolerancing machined parts. Before this method was developed, long and often confusing notes were used to describe part features. Now, with geometric dimensioning and tolerancing, these complicated notes are replaced with easy-to-read symbols. In addition to showing the size and location of part features, this method also shows exactly how far part features can vary from true form and still function as intended.

Geometric dimensioning and tolerancing is normally used to control three part characteristics: form, location, and runout. A series of standard symbols indicates each variation. Almost any geometric form can easily be controlled with geometric dimensioning and tolerancing. The first step in mastering this dimensional form is to learn the terms and elements and how they are used.

TERMS AND SYMBOLS

The heart of the geometric dimensioning and tolerancing system is the proper interpretation of each symbol. Although you do not need to memorize these symbols, you should be able to identify each symbol and its correct meaning.

FEATURE CONTROL SYMBOL

The feature control symbol (Figure 8-1) lists the required tolerance data. This symbol is a frame that contains the geometric characteristic symbol, datum references, and tolerance value. The exact size and entries of the feature control symbol are determined by the part features that require control.

Geometric characteristic symbols show the characteristic that is to be controlled. Each of these symbols has a specific meaning and application, as shown in Figure 8-2.

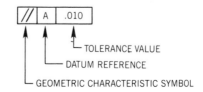

Figure 8-1 Feature control symbol.

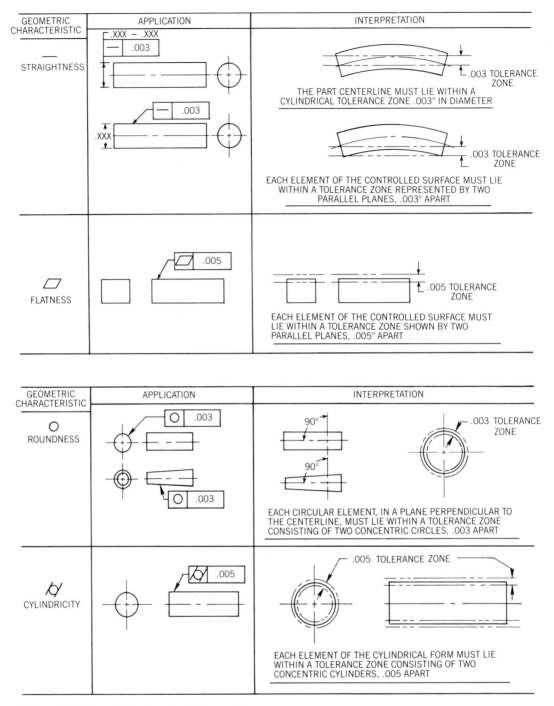

Figure 8-2 Geometric characteristic symbols.

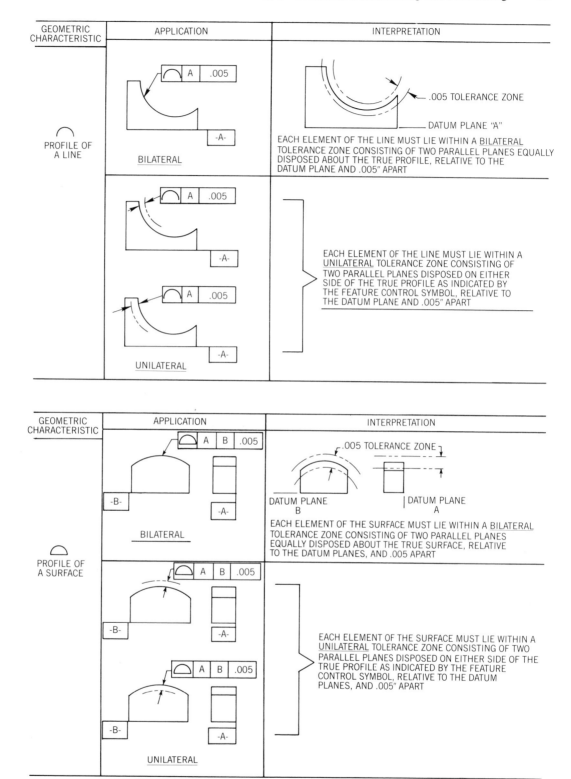

Figure 8-2 *(continued)*

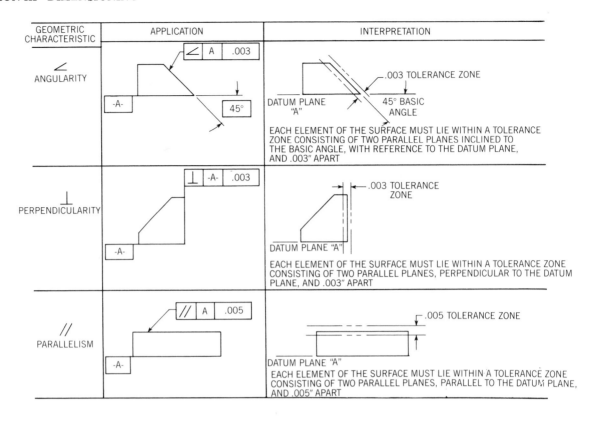

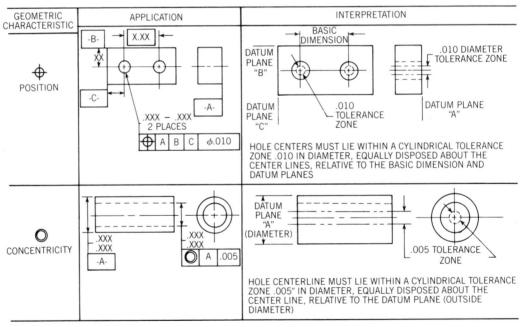

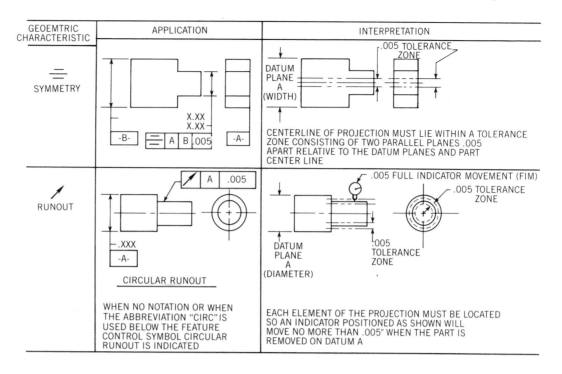

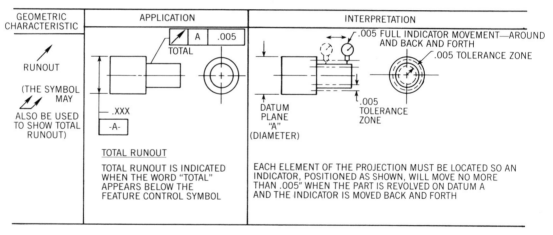

Figure 8-2 *(continued)*

Datum references are the letter values that identify a specific datum surface on the part. A datum is a line, surface, edge, or a point on the part that is used as a reference or point of origin for dimensions. The letter value shown in the feature control symbol refers to the specific datum shown with a datum identification symbol. As shown in Figure 8-3, the datum reference in the feature control symbol means the detail must be held parallel to the datum surface shown with the datum identification symbol. When more than one letter is used as a datum reference (Figure 8-4), the first letter indicates the primary datum, the second letter indicates the secondary datum, and the tertiary datum is shown with the third letter. Figure 8-5 shows how these three datums are

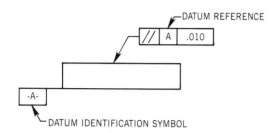

Figure 8-3 Application of datum reference and datum identification symbols.

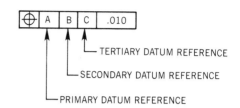

Figure 8-4 Primary, secondary, and tertiary datums.

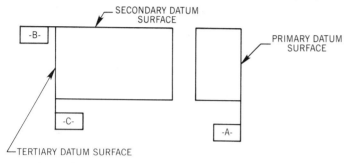

Figure 8-5 Relating datums to an actual part.

Figure 8-6 Dual primary datums.

related to a single part. In cases where a part has two datum surfaces that are primary datums, the datum references will have a dash between them (Figure 8-6).

Tolerance values show how much the part can vary from the dimensioned size and still pass inspection. Remember this is a *total* amount, not a plus or minus value.

SUPPLEMENTARY SYMBOLS

Supplementary symbols define and further clarify the meaning of the other entries in the feature control symbol.

Maximum material condition, or MMC, is the size of a part when it has the most material allowed by the tolerance. As shown in Figure 8-7, this condition means the largest size of an external feature and the smallest size of an internal feature. The key point to remember here is *most material*. The opposite condition from the MMC is the *least material condition,* or LMC. Here the part has its least material: the smallest size of an external feature or the largest

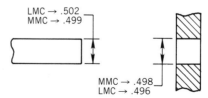

Figure 8-7 Maximum and least material conditions.

size of an internal feature. The key point to remember here is *least material*. Every part feature has both an MMC and an LMC.

The MMC modifier, indicated by a circled Ⓜ, is used only for those parts that can vary in size. When the MMC modifier is applied to a tolerance value (Figure 8-8), it means that the tolerance shown only applies when the feature is at its maximum material size. As the size of the part feature departs from the MMC size, the tolerance value increases by the amount of variation. To illustrate how this occurs, refer to the chart shown in Figure 8-9. The size of the feature is .500 ± .003, and the tolerance in the feature control symbol is .002″. The MMC size of the part is .503″, and the LMC size is .497″. As shown in the chart, the .002″ tolerance only applies at the .503 dimension. As the feature size departs from the MMC, the tolerance increases by the amount of variation.

The MMC modifier can also be applied to a datum reference if the reference is subject to a change in size. As shown in Figure 8-10, as the size of the datum varies from the MMC size, the tolerance values also vary by the amount of departure from the MMC. Only the MMC modifier is called out in the feature control symbol. The LMC is a reference value and has no modifier in the feature control symbol.

Regardless of feature size, or RFS, is a modifer indicating that the tolerance applies regardless of the size of the feature. Like the MMC modifier, the RFS modifier may be applied to either a feature size or a datum reference. Likewise, the RFS modifier, shown by a circled Ⓢ, only applies to a feature or datum that can vary in size. The effect of the RFS modifier is shown in Figure 8-11.

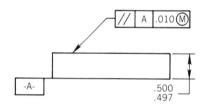

Figure 8-8 Applying an MMC modifier to a tolerance.

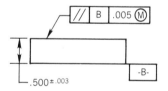

DETAIL SIZE	TOLERANCE ALLOWED
.503 MMC	.005
.502	.006
.501	.007
.500	.008
.499	.009
.498	.010
.497 LMC	.011

Figure 8-9 Interpreting the effect of MMC on a dimension.

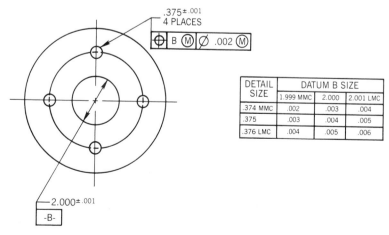

DETAIL SIZE	DATUM B SIZE		
	1.999 MMC	2.000	2.001 LMC
.374 MMC	.002	.003	.004
.375	.003	.004	.005
.376 LMC	.004	.005	.006

Figure 8-10 Applying the MMC modifier to a datum and a tolerance.

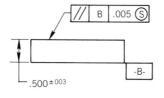

TOLERANCE	
DETAIL SIZE	ALLOWED
.503 MMC	.005
.502	.005
.501	.005
.500	.005
.499	.005
.498	.005
.497 LMC	.005

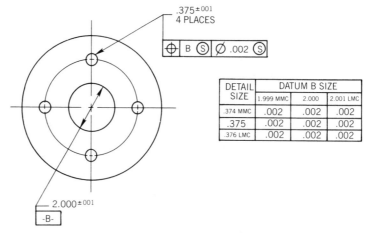

DETAIL SIZE	DATUM B SIZE		
	1.999 MMC	2.000	2.001 LMC
.374 MMC	.002	.002	.002
.375	.002	.002	.002
.376 LMC	.002	.002	.002

Figure 8-11 Interpreting the effect of RFS on a dimension and datum.

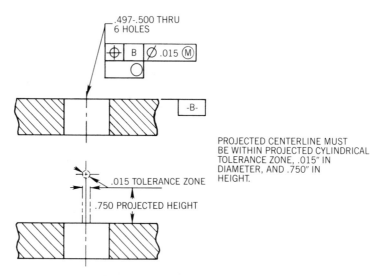

Figure 8-12 Projected tolerance zone modifier.

When the RFS modifier appears, the stated tolerance applies regardless of any variations in the part sizes.

Projected tolerance zone, shown with a circled $\textcircled{P}$, indicates that the tolerance zone is to be projected above the surface of the part (Figure 8-12). This modifier is normally positioned below the feature control symbol and is used for applications where mating parts must have aligned holes.

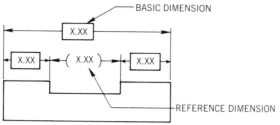

Figure 8-13 Basic dimension.

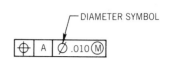

Figure 8-14 Diameter symbol used in a feature control symbol.

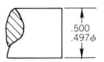

Figure 8-15 Diameter symbol used to abbreviate diameter.

Figure 8-16 Usual and international sequence of tolerance data.

Basic and reference dimensions are shown on geometrically toleranced prints by a box or parenthesis. The box (Figure 8-13) is used to indicate a basic dimension. Basic dimensions are considered to be exact. The tolerance, if any, is applied to the basic dimension. Reference dimensions are for informational purposes only and are shown in parenthesis [()].

Diameter symbols show the tolerance value applied to a diameter. In feature control symbols, the diameter symbol is placed before the tolerance value (Figure 8-14). This symbol may also be used to indicate a diameter abbreviation (Figure 8-15).

SEQUENCE OF TOLERANCE DATA

The tolerance data in a feature control symbol may be placed in either the usual or the international sequence (Figure 8-16). The only difference between these two symbols is the order of the datum reference and the tolerance value. The usual sequence places the datum reference first and the tolerance data last. The international sequence simply reverses the position of these two values.

SELF TEST

1. Identify the geometric characteristic symbols shown in Figure 8-17, a–n.

2. Identify the supplementary symbols shown in Figure 8-18, a–g.

3. Identify the elements in the feature control symbol shown in Figure 8-19, a–g.

4. Insert the allowable tolerance values in the chart in Figure 8-20.

5. What is the size of datum in Figure 8-21?

6. What is the MMC size of the feature?

Figure 8-17

Figure 8-18

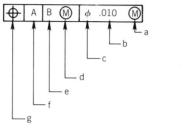

Figure 8-19

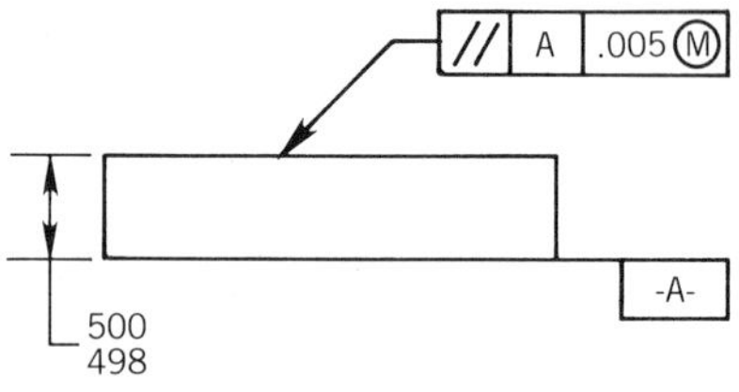

FEATURE SIZE	ALLOWABLE TOLERANCE
.500	a
.499	b
.498	c

Figure 8-20

Figure 8-21

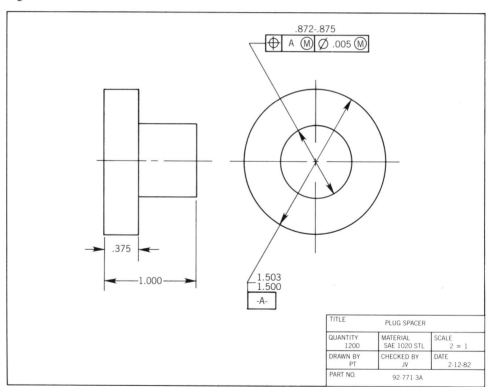

7. What is the LMC size of the feature?

8. What characteristic is to be controlled?

9. What is the tolerance of the characteristic to be controlled?

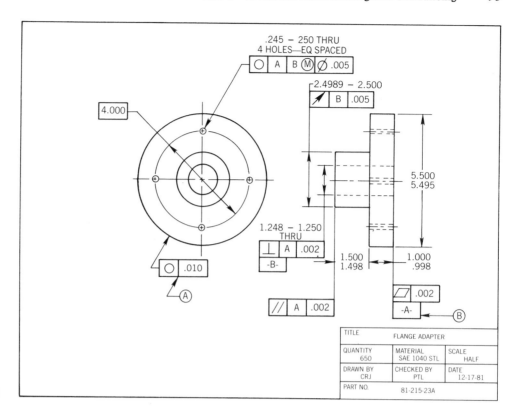

Figure 8-22

10. In Figure 8-22, what is the part name?

11. What is the part number?

12. How many parts are to be made?

13. How many datum are there on this part?

14. What type of dimension is used to show the size of the bolt circle?

15. What is the location of datum A?

16. What is the location of datum B?

17. What is meant by the feature control symbol shown at A?

18. What type of dimensions are used to show the sizes of this part?

19. What relationship exists between datum A and datum B?

20. What is the MMC of the center hole?

21. What is the LMC size of the four holes in the flange?

22. Identify the characteristic symbol used with the hub diameter.

23. What does this symbol mean with relation to the part?

24. What supplementary symbol is shown at B?

25. What does the characteristic symbol used with the size of the flange holes indicate?

**Anwers To
Self Test**

1. a. Profile of a line

    b. Straightness

    c. Parallelism

    d. Position

    e. Perpendicularity

    f. Runout (circular)

    g. Symmetry

    h. Roundness

    i. Concentricity

    j. Cylindricity

    k. Profile of a surface

    l. Flatness

    m. Runout (total)

    n. Angularity

2. a. Projected tolerance zone

    b. Reference dimension

    c. Datum identification symbol

    d. Regardless of feature size modifier

    e. Basic dimension

    f. Maximum material condition modifier

    g. Diameter symbol

3. a. Maximum material condition modifier to tolerance

    b. Tolerance value

    c. Diameter symbol

    d. MMC modifier to datum

    e. Secondary datum reference

    f. Primary datum reference

    g. Geometric characteristic symbol (position)

4. a. .005

    b. .006

    c. .007

5. 1.500–1.503″

6. .875″

7. .872″

8. Position

9. .005″

10. Flange adapter

11. 81–215–23A

12. 650

13. 2

14. Basic

15. Back surface of flange

16. Inside diameter of center hole

17. The flange must be round within .010″.

18. Limit dimensions

19. Datum B is established perpendicular to datum A

20. 1.250″

21. .245″

22. Runout (circular)

23. The runout must not exceed .005″ with relation to the center hole.

24. Datum identification symbol

25. The holes must be positioned within .005 with relation to datums A & B.

9